THE PERIODIC TABLE AND A MISSED NOBEL PRIZE

THE PERIODIC TABLE AND A MISSED NOBEL PRIZE

Ulf Lagerkvist

Gothenburg University, Sweden

editor

Erling Norrby

The Royal Swedish Academy of Sciences, Sweden

 **World Scientific**

NEW JERSEY · LONDON · SINGAPORE · BEIJING · SHANGHAI · HONG KONG · TAIPEI · CHENNAI

Published by

World Scientific Publishing Co. Pte. Ltd.

5 Toh Tuck Link, Singapore 596224

USA office: 27 Warren Street, Suite 401-402, Hackensack, NJ 07601

UK office: 57 Shelton Street, Covent Garden, London WC2H 9HE

British Library Cataloguing-in-Publication Data
A catalogue record for this book is available from the British Library.

ISBN 978-981-4295-95-6 (pbk)

Typeset by Stallion Press
Email: enquiries@stallionpress.com

Printed in Singapore by Mainland Press Pte Ltd.

Contents

Foreword

In May 2010 the Sven and Dagmar Salén Foundation decided to give a grant to Professor Ulf Lagerkvist to allow publication of a manuscript entitled *The bewildered Nobel Committee* by the World Scientific Publishing Company (WSPC), Singapore. This decision was based on a thorough review by Torbjörn Norin, professor of organic chemistry at the Royal School of Technology in Stockholm and a member of the board of the foundation at the time. Lagerkvist unexpectedly died a month later on June 30, 2010. I knew him by the professional contacts we had at the Royal Swedish Academy of Sciences. He was a very likable person and he knew his science. With time he also documented himself to be a highly competent author presenting excellent books popularizing the history of science. This talent might not have come as a surprise since his father was Pär F. Lagerkvist, who in 1951 had received the Nobel Prize in literature "for the artistic vigour and true independence of mind with which he endeavours in his poetry to find answers to the eternal questions confronting mankind."

In order to see if there was a way to resurrect the manuscript I searched the archives of the Salén foundation, but the text was not to be found. In September 2011 I was visiting WSPC, which a year previously had published my book *Nobel Prizes and life sciences* (1). Contacts with Assistant Director Joy Quek revealed that she had managed to retrieve the manuscript from Ulf Lagerkvist's daughter Louise Lagerkvist Olemyr, who in turn had found it in her father's computer. Thus it was possible to save the text and what remained was only to select pictures to accompany the text. It was a pleasure to take responsibility for this. The task was facilitated when I discovered that Lagerkvist already in February 2009 had some preliminary contacts, inquiring about picture material, with archivist Anne Miche de Malleray at the Center for the History of Science at the Academy, where I have my office. Lagerkvist's text has been used in this book essentially without modifications except for some minor adjustments based on suggestions from Norin and also from my cousin Lars-Johan Norrby, professor emeritus of inorganic chemistry at the Stockholm University. The latter also kindly helped me to select the pictures that were eventually used.

It deserves to give some additional background to Ulf Lagerkvist's life journey. He and his twin brother Bengt were born in 1926. They were the only two children of his father's second marriage. The two sons went different ways with Ulf becoming a biochemist and during the later part of his life author and Bengt establishing himself, after having received a law degree, with considerable success as a theater director and television producer. He has also been responsible for very much appreciated programs on art and artists on Swedish television. Apparently the two brothers had very close contacts and after Ulf's death Bengt wrote a richly emotional obituary emphasizing the unique relationship that may exist between twin brothers.

Ulf's career started at the Karolinska Institute, where he received his basic education. He did not finish his medical studies, but instead he developed into a researcher in biochemistry under the guidance of the nucleic acid chemist Einar Hammarsten. Hammarsten was very influential in the Nobel committee at the Karolinska Institute for many years and he has been presented in my previous book (1, chapter 7). Ulf's work focused on the metabolism of the components of the nucleic acids and he made important contributions in this field. As a young scientist he was recruited to the Medical Faculty of the Gothenburg University where he developed his talent for science in a powerful way. In the beginning of the 1960s he was a visiting scientist in the laboratory of Paul Berg at Stanford University, Palo Alto, California, who later became the recipient of the 1980 shared Nobel Prize in chemistry. This visit came to further establish his position as an influential scientist in the rapidly developing field of molecular biology. The various roles of the nucleic acids in protein synthesis and the mechanisms for a correct reading of the genetic code were in the focus of his studies. In 1974 he was appointed professor and chairman of medical and physiological chemistry at Gothenburg University. He became a member of the Royal Swedish Academy of Sciences in 1983 and was an important contributor as a reviewer in the work by the Nobel committees for chemistry for many years.

After his retirement in 1991 Lagerkvist shifted towards developing his talent for writing. He became a much respected author by the series of books he published over the years. In the beginning he wrote in Swedish, but later he switched to English, a language of which he had a considerable command. He wrote about his father in 1991 — *Den bortvändes ansikte: en minnesbok* (The averted face: a memory book) and much later, in 2010, about his mentor *Människofiskaren* (The fisherman of humans) *Einar Hammarsten*. He also wrote about science policy publishing in 1999 perspectives on the Karolinska Institute and its fight against the Swedish universities. But in particular he became engaged in popularizing science, starting in Swedish with *Gener, molekyler, människor* (Genes, molecules, humans) and then switching to English; *DNA Pioneers and their Legacy* (2), *Pioneers of Microbiology and the Nobel Prize* (3) and *The Enigma of Ferment* (4). The first of these books carried a dedication "To the memory of Einar Hammarsten who introduced me to nucleic acids." Lagerkvist's last manuscript resulted in this book, but in the editing process it was decided to change the title to *The Periodic*

Table and a missed Nobel Prize. I hope he would have accepted this modification although throughout his text he preferred to talk about the periodic system.

The exceptional prestiges of the Nobel Prizes in natural sciences rely on the fact that the discoveries that have been recognized in a historical perspective have been confirmed to be milestones of development. In this context it is sad that the discovery of the periodic table never was recognized by a prize. The reason that the Royal Swedish Academy of Sciences missed the opportunity to award the Russian Dmitri Mendeleev in 1906 reads like a detective story. The outcome was decided by one single academy member, Professor Peter Klason! At that time the committees were still struggling with the formulation in Alfred Nobel's will which said that a discovery made during the "previous year" should be recognized. Still a modification of the statues already at this time did allow recognition of a discovery made earlier, provided the insight into its importance had become obvious only recently, sometimes because of the emergence of addition information. The full appreciation of the central role of the periodic table — the "Rosetta stone" of chemistry — did indeed require supplementary data. Elements predicted by Mendeleev to exist were discovered and the existence of one additional group of elements, not known by him, was identified and was later awarded the Nobel Prize in chemistry in 1904. This prize was given to William Ramsey "in recognition of his services in the discovery of the inert gaseous elements in air, and his determination of their place in *the periodic system* (my italics)". In the final discussions Klason, tactically, in fact acknowledged the importance of this added knowledge, only to skillfully hereafter pull another card out of his sleeve. But it is now time to let Lagerkvists's text speak for itself. Suffice to note that in 1907 both Mendeleev and Henri Moissan, who did receive the 1906 Nobel Prize in chemistry, both died. It could also be added that Mendeleev and Nobel did finally meet, albeit posthumously. In the present form of the periodic table the synthetic elements include as number 101, Mendelevium (*Md*) and as its nearest neighbor, number 102, Nobelium (*No*).

Erling Norrby

1. Norrby, Erling. *Nobel Prizes and Life Sciences* (World Scientific, Singapore, 2010).
2. Lagerkvist, Ulf. *DNA Pioneers and Their Legacy* (Yale University Press, 1998).
3. Lagerkvist, Ulf. *Pioneers of Microbiology and the Nobel Prize* (World Scientific, 2003).
4. Lagerkvist, Ulf. *The Enigma of Ferment* (World Scientific, 2005).

Preface

When the King of Sweden on 10 December each year awards the Nobel Prize to a carefully selected group of scientists, this solemn occasion is the result of the great efforts of many hands. Clearly the most delicate task of all is the choice of the Nobel laureates from the hundreds of candidates nominated. To give an example, the number of nominees for the prize in chemistry increased from 270 in the year 1994 to 350 in 2006. Much of this work falls on the Nobel Committee for Chemistry of the Royal Swedish Academy of Sciences. The number of people making up this body has varied over the years. In 1906 the committee had five members while a century later it contained eight. To assist it the committee can call on a considerable number of experts in the different fields of chemistry who will submit written reports on the candidates that have been assigned to them. Eventually the Nobel Committee reports its findings and conclusions to the Academy where the report is first scrutinized by the chemists of the Academy, who make up the so-called class of chemistry. The final decision is taken by the whole Academy in plenary session.

It is a long and fairly complicated process before the prize and its insignia can at last be handed over by the King and it is not to be wondered at that differences of opinion and conflicts can occur on the way. If again we take the prize in chemistry as an example, the members of the Nobel Committee for Chemistry have often enough disagreed with each other in the past and sometimes the debates have been heated. The most well-known example of this occurred in 1906 and the results are discussed to this day. In the present book I will attempt to relate and analyse the course of events which took place a century ago and concerned one of the greatest breakthroughs in theoretical chemistry, the discovery of the periodic law, and whether that could be rewarded with a Nobel Prize. Obviously the number of candidates both in physics and chemistry who have made contributions which deserve this accolade will always be far greater than the number of laureates that can be rewarded each year. Sins of omission on the part of the Academy of Sciences, which is responsible for the Nobel Prizes in Physics and Chemistry, are consequently inevitable and therefore to some extent excusable.

The book is intended for the layman reader and does not presuppose any profound knowledge of either chemistry or physics. However, in order to discuss in their proper context the different subjects raised, it has been necessary to outline briefly some aspects of the scientific development from ancient times to the previous turn of the century when the Nobel Prizes came into being. This has been the subject of several previous books by the present author but some of these publications are only available in Swedish.

I am greatly indebted to the Royal Swedish Academy of Sciences and its chemical Nobel Committee for generously giving me access to their archive with information on Dmitri Mendeleev and Henri Moissan. I also want to thank Anne Miche de Malleray of the Center for History of Science of the Royal Swedish Academy of Sciences for valuable help with the illustrations.

Finally, I gratefully acknowledge the generous financial support of the Sven and Dagmar Salén Foundation in the publication of this book.

Ulf Lagerkvist

ELEMENTS, ATOMS AND MOLECULES

Atoms as a Philosophical Concept

The early Greek philosophers had used the term "element" to signify earth, water, air and fire, concepts encountered already in the old Egyptian world of ideas. However, eventually the word "element" took on a more definitive, chemical meaning with the introduction of the idea of elementary particles by the ancient Greek philosophers Democritus, who lived in the 5th century B.C., and Epicurus (341–270 B.C.). Their philosophy, as far as the physical world was concerned, centred on the concept of the atom. This word, derived from the Greek "*atomos*", meaning indivisible, they used to denote the smallest parts that make up matter.

The teachings of Democritus and in particular Epicurus inspired the Roman poet Lucretius (early half of the last century B.C.), who in his only extant poem *De rerum natura* (On the Nature of Things) gives a detailed account of the Epicurean philosophy, born of a deep conviction. We know practically nothing about the life of Lucretius, except a possibly false tradition to the effect that he became insane after having drunk a love potion and then committed suicide. Be that as it may, in his great poem he asserts that matter is made up of indestructible atoms so that, consequently, nothing can arise out of nothing, and nothing can be reduced to nothing; ideas which presumably originated with Democritus. The universe is infinite, i.e. an infinite number of atoms exist in a limitless void. Atoms differ in shape, size and weight and they are in constant motion and move with a velocity greater than that of light. All things in the universe, including all living organisms, are made up of atoms and void. Upon the death of the organism the atoms of which it consists become free and being indestructible they continue to exist. This is true also of the soul, which is made up of a special kind of particularly fine atoms. However, there can be no eternal life for the soul as such, in spite of the fact that its atoms are indestructible.

In Lucretius' account of the philosophy of Epicurus the gods undoubtedly exist, but they are also made up of atoms. They are remote beings and take no interest in the world and the creatures that inhabit it, nor have they created the universe. They do not watch over human beings or interfere in their lives, nor do they punish them after death. This is a very important point for Lucretius, since he wants to free man from the fear of the gods. The ideas of Democritus and Epicurus, as expounded by Lucretius, about the universe as an infinite number of atoms in a limitless void, continued to fascinate certain philosophers and to provoke the Catholic Church.

The Cardinal and the Heretic Monk

When Lucretius' great poem, having been lost for over a millennium, was rediscovered by Italian humanist Poggio Bracciolini in 1417, it fell into the hands of a remarkable prince of the Church, Nicolas of Cusa, or Cusanus as he is often called. He was born in the Rhineland in 1401 and had been given a very thorough education including, in addition to theology, studies in mathematics, jurisprudence and the humanities at the famous university of Padua, before he entered the service of the Catholic Church. Cusanus can be said to have led a double life. During his long and exceptionally successful career that would result both in a see and finally a cardinal's hat, he never seems to have questioned the absolute authority of the pope, whose loyal servant he always remained. However, at the same time this prince of the Church was a philosopher and a mystic who in his writings expressed a pantheistic faith and a belief in a limitless universe in the spirit of Lucretius. What is so remarkable is that he was able to keep these two apparently irreconcilable sides of his being completely separate from each other. The mystic and the prince of the Church never seem to have interfered with each other, although they were united in the same person.

On the moral plane we find the same duality. As a philosopher he always pleaded for the most extensive tolerance, but in his capacity as a high papal dignitary he could demonstrate both resolution and even severity when it came to upholding the interests of the Church. This moral duality was perhaps a prerequisite for his ability to conduct his philosophical writings, where the opinions expressed must have seemed offensive, at least in the eyes of Catholic orthodoxy. Doubtless it was Cusanus in his role as a prince of the Church that protected the heretic philosopher from an intrusive inspection by the watchful inquisition. He was simply much too valuable as a support for the papal policy to be sacrificed on the altar of fanatical bigotry. Cusanus died in 1464 and more than a century later another man of the Church, who stood on one of the lowest rungs of the ecclesiastical ladder and completely lacked the political ability and conciliatory personality of Cusanus, would learn in a horrible way how dangerous his ideas could be.

Giordano Bruno has become a symbol of the free and independent scientist, who incessantly seeks the truth regardless of the threats and persecutions by the powers that be, which feel in duty bound to uphold the officially established dogmas. Nevertheless, he was a mystic and a poet rather than a scientist, even if he was very much concerned with questions about the nature of the universe and other problems that we would consider belonging to astronomy and cosmology. However, it was the philosophical and religious aspects of his cosmology that led to the fatal conflict with the inquisition. To some extent one must agree with his persecutors here. It was indeed the religious consequences of his cosmology that were the most important to Giordano Bruno himself. He truly loved the idea of an interminable universe in much the same way as a religious believer loves his god. In the end he was even prepared to suffer death as a heretic at the hands of the inquisition.

***

In Campania, not far from Naples, in the ancient town of Nola, Filippo Bruno was born in 1548 as the son of a local watchman (Fig. 1). Because of the boy's obvious talents his parents sent him to school in Naples, which meant a very considerable economic sacrifice. At the age of 17 Filippo entered the monastery of the Dominican Order in Naples and as dictated by an old tradition he changed his name from Filippo to Giordano to signify his new position as a novice. With his inability to compromise and his generally oppositional attitude, Giordano was exceptionally ill suited for the life of a novice and he would later give a bitter account of life in the monastery in his play *Il Candelaio* (The Torchbearer). On the other hand, the Catholic Church was really the only career that was open in Italy at this time to a young, talented man completely without influential relatives or economic resources. A number of popes had started in similar humble circumstances and if it had only been a question of intellectual talents Brother Giordano might very well have attained a high office in the Church. However, this was not to be. After 11 years as a monk Giordano Bruno had got into a hopeless conflict with the authorities of the abbey. Among the serious accusations against him was that of having concealed a prohibited book by the great humanist Erasmus of Rotterdam in the privy of the abbey. In the end Giordano decided to run away from the monastery and at the same time he shed his habit as a Dominican. This would prove to be a fatal decision and in spite of several attempts on the part of Giordano Bruno to reconcile himself with the Dominican Order they continued to regard him as a runaway monk and a heretic.

Having escaped from the monastery, Bruno spent 16 years travelling all over Europe, constantly looking for patrons willing to support him economically. He was an expert in mnemonics, a technique for aiding the memory by connecting a series of unrelated ideas into an artificial whole, for instance a verse for remembering the number of days in a month. Mnemonics was extremely popular at this time and Bruno was in great demand because of his expertise. Even the French King Henry III became interested and for two years Bruno stayed in Paris and lectured at the Collège de France. Another patron was the French ambassador in London, Michel de Castelnau, and the time that Bruno spent in England during 1583–1585 was probably the happiest of his restless life and it was here that he

Fig. 1.   Giordano Bruno (1548–1600).
The statue adorns the Campo dei Fiori in Rome. It was unveiled in 1899 despite objections from the Pope Leo XIII.

wrote his first major philosophical and cosmological books. Having returned to Paris, Bruno soon became embroiled in quarrels with local scholars and he continued his incessant travelling in Germany, Switzerland and Northern Italy. A Venetian nobleman, Giovanni Mocenigo, invited him to Venice. He wanted Bruno to teach him mnemonics, but teacher and pupil soon fell out and after a stormy scene Mocenigo saw to it that Bruno was arrested in the middle of the night in May 1592 and thrown into the dungeons of the inquisition. He was brought before a tribunal that finally managed to extract a confession from the prisoner in which he abjured all heresy and begged to be reconciled with the holy Catholic Church.

While this was going on in Venice the inquisition in Rome became interested and asked to have Bruno delivered into its hands. The Venetian authorities were reluctant to agree to this request, which they saw as an infringement on their own province, but when the pope, Clement VIII, intervened they gave in and Bruno was in February 1593 delivered to the Roman inquisition in whose dungeons he was to remain until he was executed seven years later.

What was the nature of his unforgivable heresy? Bruno's philosophy and cosmology is about the interminable universe, an idea that he had argued in a number of books from 1584 until he got into the clutches of the inquisition in 1592. But how original are his concepts? In reality his limitless universe with its countless celestial bodies made up of innumerable atoms is no different from the cosmos envisaged by Cusanus and Lucretius. Furthermore, a contemporary of Bruno, English astronomer Thomas Digges, had already in 1576 proposed an interminable universe. We know that Bruno had thoroughly studied the work of Cusanus and that he carried with him on his countless travels a copy of Lucretius' great didactic poem. Nevertheless, there are certain differences between Bruno and his predecessors, but they are more a question of presentation than of actual cosmological facts. Where religion was concerned, Lucretius was completely neutral; he did not deny the existence of gods but he was not really interested in religion. His moral ambition was to free humanity from its fear of the gods. His disciple Cusanus, on the other hand, was in every way a Christian and he always endeavoured to harmonize his own cosmology with the demands of the holy Catholic Church. This was in contrast to Giordano Bruno who certainly did not accept the views of Catholic dogmatism with its narrow-minded intolerance that he always rejected. Sometimes he used very provocative language, for instance when he claimed that Italy was "crushed under the feet of the abominable priests". On the other hand, there is a certain pantheistic religious keynote in his cosmology, which sets him apart from the rationalism of his master Lucretius.

During his long imprisonment Bruno was repeatedly brought before tribunals of the inquisition but to this day it remains something of a mystery what it was in his somewhat nebulous cosmology that was so threatening to the Church. After all,

Cusanus had said pretty much the same thing a century earlier without the inquisition taking any action. Maybe it was really a question of Bruno's personality, his sullen obstinacy and uncompromising refusal to submit to the authority of the Church that was the reason for his misfortunes. Eventually he was sentenced to burn at the stake and on 17 February 1600 the execution took place in Campo dei Fiori in Rome.

Bruno's cosmology is not original; he had a number of predecessors like Epicurus, Lucretius, Cusanus and Thomas Digges. However, he was an uncompromising and fearless rebel to the very last and it is as a symbol of free thought that he has his greatest importance. This is why the pyre in Campo dei Fiori has been a beacon of humanity for all times.

The Dawn of Chemistry

The ideas of Democritus and Epicurus continued to be important concepts in philosophy but at the same time they were of little consequence for experimental chemistry. It would not be until this science matured during the scientific revolution in the 17th and 18th centuries that the idea of matter being made up of atoms became a fundamental principle of chemistry.

Primitive chemistry must have originated with the use of fire in the cave of Neanderthal man in order to prepare food and make it tastier. Thus, chemistry is a science that in all probability was invented by women when they by trial and error learnt how to treat the vegetables they gathered and the raw meat that the men brought home from the hunt. With the transition from hunting and gathering to a life based on agriculture, the demand for more sophisticated and efficient tools led to the mining of ores and to procedures for the melting and processing of metals. The interest in chemistry changed from the preparation of biological material to make it suitable as food and was instead focused on metallurgy and the forging of metal hardware and weapons. What had started as a female activity became a science for men and remained that way until the last century, when women began to reclaim what was once their preserve.

At the time when man first made fire his servant and began to perform simple chemical operations it was a decidedly practical activity and probably did not involve any theoretical considerations. This may have changed with the advent of a higher civilization that involved the use of metals. Sometime during this period of increasing familiarity with metal work, a theory was developed that came to be known as alchemy, which can with some justification be regarded as the first comprehensive chemical theory. In fact, the word "alchemy" is probably derived from the Greek "*cheo*", meaning "to pour or cast", referring to metallurgic activities. The earliest written documents dealing with alchemy are from the 3rd century A.D., but there is reason to believe that this cross between science and pure mysticism goes as far back as the Hellenistic culture that had its centre in Alexandria during the last centuries B.C.

Some of the basic ideas of alchemy can be traced to the leading Greek philosopher Aristotle (384–322 B.C.) (Fig. 2) whose theories about nature dominated scientific thinking for almost two millennia after his death. Aristotle was of the opinion that all substances of which the earth is made up are ultimately derived from a "prime matter" that can take the form of the four elements. By combining with each other in certain proportions under the right conditions the elements can give rise to all material objects. As a consequence of this theory of a prime matter from which everything originates, it was surmised that by suitable manipulations all substances could be transformed into each other. In ancient Egypt there had since time immemorial existed a class of artisans highly skilled in metallurgy. Inspired by the ideas of Aristotle and drawing on their own experience with alloys that attempted to imitate gold, they could have hit on the idea that it might be possible to find conditions where base metals such as lead and copper would be transformed into real gold.

Thus, it seems reasonable to assume that the roots of alchemy as a practical experimental activity in the laboratory go back all the way to such Egyptian metal workers. In any case, as we shall see in the following, alchemy represents the first attempt to build laboratories equipped with a source of heat as well as instruments and vessels that made it possible to perform real chemical experiments. This is a great and decisive step from the unbridled speculations of Aristotle's philosophy to the modern view that the experiment is paramount in research. There was no shortage of fanciful speculations in the heyday of alchemy, but its practitioners at least recognized the desirability of experimental verification.

The writings of the alchemists from the early centuries A.D. that have come down to the present time do not make easy reading. First of all, the artisans that dealt in gold making had no real wish to make their methods known to possible competitors and they therefore invented a complicated terminology, not to say a secret language, which made their texts next to incomprehensible. Furthermore, alchemy became increasingly a haven for all kinds of mysticism, which tended to obscure its practical experimental inheritance from Egyptian metallurgy. Nevertheless, in spite of all these difficulties of communication, alchemy was widely disseminated in all the great civilizations, from China to the rapidly rising Arab empire that by the 8th century stretched from Spain to the Indian subcontinent. The Islamic world with its more tolerant attitude contained a number of refugees, who had previously fled from persecutions in their homelands to

Fig. 2.   Aristotle (384–322 BCE).
Picture of statue from the Internet.

countries later conquered by the Arabs. Here we find for instance the Nestorians, who had been expelled from Byzantium in the 5th century because they denied the divine nature of the Virgin Mary. With these numerous learned refugees alchemy reached the Muslim countries, which turned out to provide a very fertile soil for it.

The Arab conquerors were deeply interested in Greek learning, particularly in the fields of medicine and science and they had a number of manuscripts on these subjects translated from the Syriac of the Nestorians to the Arabic language. Among these were several that dealt with alchemy and for centuries the Islamic caliphates, both in Baghdad under the Abbasids and later also in Egypt under the Fatimids and in Spain, became centres of alchemy. Such leading names in Islamic medicine as Rhazes (c. 865–c. 920) and Avicenna (980–1037) undoubtedly took an interest in alchemy, but they seem to have been mainly fascinated by its putative ability to cure disease. When it came to the chemical operations in alchemy the most important author was Jabir ibn Hayyan, also known under his Latinized name Geber, who according to tradition lived in the 8th century. However, although Jabir is said to have been the author of a number of books in alchemy, in all probability most of them have just been ascribed to him because his great fame as an alchemist reflected credit on them. Among the numerous books that he supposedly wrote there are four, one of them with the title *Summa Perfectionis*, which were translated into Latin and became very influential. Regarding these books there can be little doubt that they were probably written not by Jabir but by a Spanish alchemist at the beginning of the 14th century, who had taken the name of Geber. These Latin texts by Geber contain not only an unusually clear account of alchemical theory, but also an extensive and most important manual for chemical laboratory work. To these books we owe most of what we know about the equipment of an alchemist's laboratory, such as the furnace, the apparatus used for distillation and all sorts of vessels like flasks and beakers. Furthermore, they describe the actual procedures used to prepare different reagents, for instance mineral acids. We can safely assume that alchemy represented all the knowledge of inorganic chemistry from the Hellenistic period to the Renaissance and that it was responsible for the development of the chemical laboratory and the experimental procedures employed there. Modern chemistry can with some justification be said to have its roots in the ancient knowledge of the alchemists.

Atoms and Corpuscles

In the 17th century the term *"minima naturalia"* was often used to signify the smallest size of a compound where it was stable and kept its form. In other words, it could not be further divided without losing its specific chemical character. Thus, many authors used the term as a synonym for atoms and the new concept of corpuscles. At this time the four elements of Aristotle — water, fire, air and earth — were beginning to be replaced as the fundamental basis of matter by these newfangled notions.

Already Paracelsus (1493–1541), or Philippus Aureolus Theophrastus Bombastus von Hohenheim, to give him his proper name — that revolutionary figure in the history of medicine and chemistry — had substituted the three chemical principles of mercury, sulfur and salt for Aristotle's elements, claiming the balance of the three principles in the human body ultimately determined sickness and health. This attempt at a chemical explanation of the basic phenomena in physiology and medicine came to be known as iatrochemistry ("*iatros*" meaning "physician") and would be an important figure of thought in biomedicine for several centuries. The most famous and influential protagonist of iatrochemistry was Flemish aristocrat Johann Baptista van Helmont (1577–1644). As the scion of a noble family of considerable wealth he never practised as a physician and his interest in biomedicine was purely theoretical and philosophical. Nevertheless, his sarcastic criticism of Hippocratic medicine and its obsession with its bogeyman, the dangerous phlegm, was merciless and deeply offended his colleagues among the physicians. Perhaps even more shocking to them was his complete rejection of that mainstay of all therapy, bloodletting.

While it is easy enough to agree with van Helmont when he rejects the four body humors of Hippocrates as the all-explaining theoretical foundation of medicine and condemns bloodletting as the destroying angel of conventional therapy, the theories that he himself proposes in his *Ortus Medicinae* are equally fantastic to the modern reader. At the same time we must not forget that van Helmont made important chemical discoveries. He is perhaps best known for having introduced the term "gas" to signify for instance the compound formed by the burning of charcoal or the fermentation of must, carbon dioxide, which he called "*gas sylvestre*", meaning "wild gas", because he found it difficult to constrain. He also described a number of other gases, among them chlorine and sulfur dioxide as well as methods for the preparation of the corresponding acids, hydrochloric acid and sulfuric acid. Like all alchemists van Helmont was deeply interested in the putative transformation of metals into each other. In particular he was interested in mercury, which had been introduced as a medicine by Paracelsus, and van Helmont recommended the use of mercury preparations as a remedy for a number of illnesses. The introduction of heavy metals into the pharmacopoeia at this time must have caused untold suffering to the poor patients over the centuries.

Fermentation had been a central concept in alchemy where it was used to denote a number of chemical transformations aimed at producing gold from base metals and brought about by the influence of the philosopher's stone or sundry elixirs with the same effect. Ferments and fermentation fascinated van Helmont and he believed that all processes in the organism were caused by ferments that somehow converted the ingested food into living flesh.

Franz de le Böe, called Sylvius (1614–1672), was a very successful clinician and professor at the University of Leyden, where he reformed the medical education and introduced clinical training of the students in the wards. He made fermentation the central theme of his medical theories and believed that the breakdown of food brought about by ferments eventually produced end products that were either taken up by the blood or excreted as faeces. Sylvius' theory was that sickness and health depended on the delicate balance between the acid and alkaline end products of what would today be called metabolism. In health these products are in equilibrium but if the balance is disturbed an alkaline or acid excess may result in disease. All illnesses can therefore be classified as either alkaline or acid in nature. Thus, Sylvius' concepts of physiology and medicine could in some respects seem fairly close to our modern views, but it is important to realize that "ferment" at this time had a considerably wider and less definitive meaning than the present term "enzyme". It is tempting to over-interpret similarities in the terms and ascribe an insight to the 17th century iatrochemists that they did not possess.

Another famous professor at the medical faculty of Leyden, Hermann Boerhaave (1668–1738), was not only the leading clinician of his time but also held the chairs in both botany and chemistry. He wrote a textbook in chemistry, *Elementa Chemiae*, which became very popular and influential, where he laid great weight on exact methods and experimental work rather than fanciful theories. Boerhaave was not in favour of the speculations in iatrochemistry as expounded by Sylvius, and he used the term "fermentation" in a much more precise way. In the heyday of alchemy it had been used to denote all sorts of chemical transformations but Boerhaave limited fermentation to mean a process by which vegetable materials were transformed into alcohol and later into vinegar. This can be seen as the first attempt to give the term "ferment" a stricter implication that would eventually come close to the modern concept of "enzyme".

An Unlikely Career

The youngest son of the 1st Earl of Cork, Robert Boyle (1627–1691) (Fig. 3), achieved lasting fame as a philosopher, physicist and chemist, an unlikely career for a man with his background. In consequence of his empiricism Boyle held that the experiment was of paramount importance in science and his extensive contributions to the experimental arsenal of chemistry make him a pioneer of chemical methodology. Nevertheless, his most influential chemical publication, *The Sceptical Chymist* (1661), deals with more theoretical questions. Here he refutes not only the four elements of Aristotle but also the ideas of Paracelsus that

Fig. 3.   Robert Boyle (1627–1691).
Courtesy of the Royal Swedish Academy of Sciences, Stockholm.

mercury, sulfur and salt were principles whose proportions in the living organism determined health and disease. Instead he suggested that matter was made up of corpuscles, which constituted its smallest elementary particles, an idea that was far from original at the time. Such terms as "atoms" and "*minima naturalia*" were frequently used by chemists on the continent of Europe. However, Boyle also pointed out that by thinking in terms of corpuscles and their movements, one could understand such properties of matter as heat, fluidity and solidity. Furthermore, it was possible to explain chemical reactions by assuming that disparate corpuscles might form groups leading to chemical reactions that gave compounds with properties different from the original reactants.

The corpuscles that in Boyle's view made up matter had not only mechanical properties. He also believed that what he called semens (seeds), spirits and ferments contributed to the properties of matter. The idea of seeds and seminal principles was important to Boyle's thinking and he considered them to be a kind of corpuscles endowed with the power of fashioning particles that could ultimately give rise to living organisms. The deeply religious Boyle believed that this property was the result of God's direct intervention in nature and he consequently rejected the concept of spontaneous generation, which was widely accepted at the time. It was for instance believed that small organisms could arise spontaneously in the presence of suitable nutrients. Like van Helmont and Sylvius, Boyle was of the opinion that a number of chemical reactions took place in the living body. He therefore considered fermentation to be of the greatest importance both for physiology and pathology. In his view chemistry was the foundation of medicine.

Boyle did not always distinguish clearly between such concepts as *minima naturalia*, atoms and corpuscles. These terms were used interchangeably to denote the smallest particles into which natural bodies can be decomposed. However, one gets the impression that corpuscles represent a higher organization of atoms and *minima naturalia*, being to a higher degree endowed with chemical and perhaps even biological properties.

In France and other countries on the continent of Europe a number of chemists had adopted corpuscular ideas at this time. In most cases corpuscular and iatrochemical notions were closely linked to each other. However, the great philosopher Descartes (1596–1650) had explained both geology and chemistry as strictly mechanical phenomena. For instance, the particles of freshwater were soft and flexible while those of salt water were hard and rigid. The formation of different kinds of salt was the result of a purely mechanical process. Descartes believed that common salt, by changing the form of its crystals, could be transformed into saltpetre. In his natural philosophy he did not accept chemistry as an independent science. Instead everything had a purely mechanistic explanation. The form of a particle determined its chemical properties. Thus, acid compounds were made up of pointed particles. In fact, as far as chemistry was concerned, the views of Descartes and Boyle were as different as could be.

There was no lack of fanciful chemical theories in the 17th and 18th centuries. Take for instance the ideas about combustion put forward by German physicians Georg Ernst Stahl (1660–1734) and Joachim Becher (1635–1682). They suggested, without any experimental support whatsoever, that when combustible material burned something was lost to the surrounding air. Stahl never clarified if this disappearing matter was actually a substance possessed of properties such as weight or should rather be regarded as the principle of combustibility. He called it phlogiston and the phlogiston theory became immensely influential and would warp the thinking of chemists for almost a century. The fact that a metal oxide, called a calx in those days, increased in weight compared to the metal being oxidized was well known to the adherents of the phlogiston theory, according to which both oxidation and combustion were explained as a loss of phlogiston to the air. Nor were they impressed by the findings of John Mayow (1641–1679) who during his all-too-short scientific career clearly demonstrated that a candle burning in a closed bell jar consumed something in the air (he called it *spiritus nitro-aereus*) that was necessary to sustain the burning of the candle. In the same vein, an animal breathing in a closed vessel consumed *spiritus nitro-aereus* and when all of it had been used up the animal would eventually die. Mayow also realized that when a metal such as antimony was transformed into its calx, it increased in weight because something (*spiritus nitro-aereus*) was taken up from the air. It would take 100 years before the importance of Mayow's findings was realized and the phlogiston theory was swept away by Lavoisier and his antiphlogistic chemistry.

A Chemical Revolution

Even outstanding chemists like Scheele and Priestley (see later) had been prisoners of the upside-down phlogiston concept but with the arrival of the truly revolutionary antiphlogistic chemistry of Antoine-Laurent Lavoisier (1743–1794) (Fig. 4) this Alice-in-Wonderland kind of chemical thinking came to an end. Lavoisier was the scion of a wealthy family and had received a broad education, which included the humanities and mathematics as well as the sciences. His mind was early set on a scientific career and he aimed at becoming a member of the prestigious French Academy of Sciences. To this end he presented several scientific papers to the Academy and at the early age of 25 he was elected to that august body as a supernumerary member, no doubt helped along by the influence of his wealthy family. He also made a fortunate marriage to the young Marie Paulze, who became not only his wife but also a devoted collaborator in his laboratory at

Fig. 4. Antoine-Laurent Lavoisier (1743–1794).
Courtesy of the Royal Swedish Academy of Sciences, Stockholm.

the Paris Arsenal where he had been made a director. Everything really looked rosy and promising for the young academy member.

He became increasingly interested in the chemical composition of the air. As discussed earlier it had been known for a long time that many metals gained in weight when they were transformed into their corresponding calxes by the putative loss of phlogiston. Lavoisier, however, became increasingly convinced that when a metal was oxidized it did not lose phlogiston but instead took up something from the air. The question then became if a metal oxide could liberate the component it had presumably taken up from the air during its formation, provided it was heated to a sufficiently high temperature. In 1774 Joseph Priestley visited Paris and Lavoisier was made aware of the fact that Priestley had produced a gas that would later be called oxygen by heating mercuric oxide. Furthermore, Carl Wilhelm Scheele had at the same time written to Lavoisier and reported that a gas, which very effectively sustained combustion, could be obtained by heating silver carbonate. However, both Priestley and Scheele were completely captivated by the phlogiston theory and misunderstood the chemical properties of the gas, which Priestley even called "dephlogisticated air" to stress its relation to the phlogiston theory. Lavoisier, on the other hand, realized the true nature of the gas. It was the component in the air that was taken up by a metal that was oxidized, as well as by a burning candle and a respiring animal. It was therefore identical with the *spiritus nitro-aereus* of John Mayow, but Lavoisier was probably unaware of Mayow's discoveries almost a century earlier.

Lavoisier had previously studied the burning of phosphorus and sulfur in the air which formed the corresponding acids and he became convinced that the burning substances had taken up the new gas that had thus become part of the acids formed (phosphoric and sulfuric acid). In any case, there was no doubt that the acids had increased in weight compared to the mother substances, phosphorus and sulfur. In 1781 he published a paper about the general nature of acids in which he suggested that the gas obtained by heating metal oxides was really the acidifying principle itself. He proposed to call it *"principe oxigène"*, which would finally become "oxygen" (the begetter of acids). This was of course a mistaken generalization by Lavoisier. Already at the time it was known that some acids, for instance hydrochloric acid and hydrofluoric acid (discovered by Scheele), did not contain oxygen.

The discovery of oxygen and its role in combustion and respiration led Lavoisier to definitively refute the phlogiston hypothesis in his paper *"Réflexion sur le phlogistique"* published in 1786. Many conservative chemists rejected the new antiphlogistic chemistry, including Joseph Priestley who remained a steadfast believer in phlogiston until his death. In 1789, the same year as the outbreak of the French Revolution, Lavoisier published his *Traité élémentaire de chimie* (Elementary Treatise of Chemistry), which he himself often referred to as marking a revolution in chemical thinking. Perhaps the most interesting part of the book deals with what we now call biochemistry, for instance the fermentation of sugar to alcohol. Lavoisier particularly emphasizes that in this process the elements oxygen, hydrogen and carbon present in the sugar appear also in the products, i.e. alcohol and carbon dioxide that result from the fermentation. Nothing is in fact created or lost in this reaction; it is just a question of alterations and modifications.

Lavoisier also took up the problem of respiration. He maintained that this process involved uptake of oxygen from the air and was similar to the combustion of carbon. Consequently, it must account for the production of animal heat. In a famous series of experiments where a guinea pig was placed in a calorimeter, he and his collaborator Pierre Laplace demonstrated that the heat produced by the animal was roughly comparable to the heat obtained by burning a piece of carbon, with the formation of the same amount of carbon dioxide as that given off by the guinea pig. However, there was a small but consistent difference in that the guinea pig produced somewhat more heat, for a certain amount of carbon dioxide formed, than was given off by the burning coal. In the same vein, the animal produced less carbon dioxide than was expected from the amount of oxygen consumed. This led Lavoisier to suggest that some of the oxygen was consumed in the formation of water from hydrogen and that the excess heat was liberated in this process.

While these pioneering investigations of energy metabolism were under way in the early 1790s, the French Revolution proceeded inexorably on its road toward unrestrained Terror. Lavoisier's father had shortly before his death purchased a title of nobility and in 1775 his son inherited the title. Another fatal circumstance was that Lavoisier had a position in the so-called Ferme Générale, which formed part of the French fiscal system, a generally hated authority. He became the target of attacks from radical journalists like the notorious Jean-Paul Marat, was removed from his position at the Paris Arsenal and arrested together with all other members of the Ferme Générale. On 8 May 1794, they were brought before the Revolutionary Tribunal, summarily convicted and executed the same day. Of this mindless judicial murder the great mathematician Joseph-Louis Lagrange sadly remarked: "It required only a moment to sever that head, and perhaps a century will not suffice to produce another like it."

The 18th century is the period when chemistry advanced from being just a collection of recipes to become a science in its own right. There are several important actors here but Lavoisier's introduction of antiphlogistic chemistry must certainly be regarded as one of the most revolutionary events in the history of chemistry.

An Atomic Theory in the Romantic Era

It is something of a paradox that it was during the Romantic period of tearful sentimentality in the arts (not to mention its somewhat bizarre natural philosophy, which can be seen as a rejection of the Enlightenment's strict rationality and worship of human reason) that for the first time an atomic theory of chemistry was presented.

Proportions in Chemistry

Stoichiometry is a central concept in chemistry today and it is almost difficult to believe that it has been only about 200 years since it was first introduced by a one-time enlisted soldier in the engineering corps of the Prussian army by the name of Jeremias Benjamin Richter (1762–1807). When after seven years' service he left the army and took up studies at the University of Königsberg he came under the influence of the great German philosopher Kant. An axiom of Kant's was that all true science is applied mathematics and Richter certainly always maintained that all chemical processes must obey mathematical laws. It is ironic that his great idol Kant believed that chemistry was not a science since its principles were only empirical. Against this Richter asserted that chemistry was particularly amenable to mathematical analysis since its main problem was to determine the proportions in which elements and substances interact with each other in a chemical reaction. Consequently, his dissertation dealt with the use of mathematics to analyse chemical processes. Richter never held an academic position but eventually became a chemist at the Royal Porcelain Works in Berlin where he remained until his death from tuberculosis at the age of 45. In the years 1792–1794 he published his magnum opus, *Anfangsgründe der Stöchiometrie: oder, Messkunst chemischer Elemente* (Introduction to Stoichiometry or the Measurement of Chemical Elements). Richter is a good example of the many scientists who have made fundamental contributions but have remained curiously anonymous to this day. His idea that the reactants in a chemical process combine with each other in certain definite equivalent proportions would turn out to be fundamental in the new chemistry that developed at the beginning of the 19th century.

***

Joseph Louis Proust (1754–1826) came from a family of pharmacists and it was therefore natural that he should take an interest in analytical chemistry, having been trained as a boy in his father's pharmacy in Angers. He completed his studies in Paris but then moved to Spain where he spent most of his scientific career. He was a professor of chemistry first in Madrid and then in Segovia at the Royal Artillery College there. In 1799 he went back to Madrid, where he was given excellent working conditions, but in 1806 he had to return to France because of conflicts between his native country and Spain brought on by the Napoleonic Wars. He spent the rest of his life in Angers where he took over the family pharmacy.

Proust's great scientific discovery, the definite proportions in which chemical elements occur in different substances, owes much to his skilful analyses of iron oxides that he published in 1794. He could show that there were two kinds of such oxides, one that contained 27% oxygen and another with 48% oxygen. Contrary to what had often been believed, there was no intermediary stage between these two kinds of iron oxides. When an oxide with a lower percentage of oxygen was transformed into one with higher oxygen content, this was always accomplished in one single step. There was never any formation of oxides with an intermediary oxygen percentage. On the other hand, the great French chemist Claude Louis Berthollet (1748–1822) had arrived at the conclusion that chemical elements could react in different proportions so that the resulting compound would show a continuum of proportions of its constituent elements (atoms). Against Berthollet's theory Proust maintained that this investigator had just analysed mixtures of two oxides, or alternatively of the metal itself and one of its oxides. In any case, there can be no doubt that by introducing the principle of definite proportions of elements in a chemical compound, Proust had made a major contribution to modern chemistry.

A Self-taught Quaker Scientist

It is remarkable that the first atomic theory of chemistry was developed by a Quaker schoolteacher who supported himself as a private teacher of mathematics and natural sciences and only for a limited time held a university position. John Dalton (1766–1844) (Fig. 5) came from a rather poor Quaker family and there is nothing in his background or upbringing that in any way portends his distinguished scientific career. He was born in provincial Cumberland and attended the local Quaker school where the teachers brought the gifted boy to the attention of a prominent member of the Quaker Society of Friends, the well-known naturalist Elihu Robinson. He was thus offered a position as an assistant in a Kendal boarding school, newly set up by the local Quaker Society of Friends. Here he had access to an excellent library and a well-equipped laboratory where he could improve his scientific education in his spare time. He also attended a number of lectures by

Fig. 5.   John Dalton (1766–1844).
Courtesy of the Royal Swedish Academy of Sciences, Stockholm.

famous naturalists and the Kendal Quaker, John Gough, took a particular interest in the boy and taught him mathematics, meteorology and botany.

John Dalton and his brother Jonathan took over the Kendal school in 1785 and obviously the Society of Friends were so impressed by his educational talents that they appointed him professor of mathematics and natural philosophy at the New College, which the Society had recently established in Manchester. This was of course a great opportunity for him — he was only 26 years old at the time — but in 1800 he resigned his professorship and opened a private Mathematical Academy in Manchester, which proved such a success that for the rest of his life he could sustain himself in this way, at the same time having ample time for his own research.

An early interest of Dalton's was meteorology and the properties of the gases which make up the atmosphere. Already in 1793 he had suggested that in a mixture of different gases, each gas acts as an independent entity. In 1801 he clearly formulated the law of the partial pressures of gases and this principle was further elaborated in his paper "On the Absorption of Gases by Water and Other Liquids" (1803). Here he raised an important question: "Why does not water admit its bulk of every kind of gas alike?" He then went on to say: "... I am nearly persuaded that the circumstance depends on the weight and number of the ultimate particles of the several gases." Thus, he seems convinced that different gases are made up of particles of different weights. In the same way, chemical reactions involve the combination of particles that differ in weight. Dalton outlined his atomic theory in *Systems of Chemistry* (1807) and *New Systems of Chemical Philosophy* (1808).

The atomic theory formulated by Dalton is built on various concepts, many of which had been around for a long time. There is the concept of chemical elements or atoms, the conservation of mass in chemical reactions (emphasized by Lavoisier, even if the thought did not originate with him), the definite proportions of elements in pure chemical compounds demonstrated by Proust, and the principle of stoichiometry laid down by Richter. Dalton now postulated that matter is made up of atoms, and that atoms of the same element have the same weight and are identical in all other properties. On the other hand, atoms of different elements, for instance hydrogen and oxygen, have different weights and differ also in other respects. Furthermore, atoms are indestructible and chemical reactions only imply that atoms are being rearranged, although they themselves remain unchanged.

To begin with, the importance of his atomic theory for chemistry was not widely realized; in fact many famous chemists such as Berthollet did not subscribe to his theories. However, over the years they became generally accepted; he became a leading member of such learned societies as the Manchester Literary and Philosophical Society and in 1822 he was elected a fellow of the Royal Society. In 1833 he was awarded a Civil List pension of 150 pounds per annum (increased to 300 pounds in 1836), a considerable sum in those days. He never married and lived for a quarter of a century at the same address in Manchester with a clerical friend, the Rev. W. Johns. He was something of a recluse whose life was filled with laboratory work and numerous lectures and who hardly ever came into society. His only vice seems to have been a weakness for an occasional pipe of tobacco.

Atomic Weights and Chemical Symbols

Jacob Berzelius (1779–1848), a leading chemist in the first half of the 19th century whom we will consider in more detail in the section "The Birth of an Academy", is known for his accurate determination of atomic weights and the symbols of the elements he introduced, which became generally accepted.

Dalton's investigations of the water solubility of different gases (measured as weight of gas dissolved per volume of water) had convinced him that gases were made up of atoms of different weights. However, his rough tables of atomic weights obviously had to be followed up by much more precise determinations. This was the problem that Berzelius addressed in his primitive laboratory. It took all his remarkable experimental talent and inexhaustible capacity for work to achieve this goal. The values of atomic weights that he determined with the unsophisticated technical means available at the time are amazingly accurate and agree closely with modern figures.

Many of the great chemists of the past had tried to develop a language of chemical symbols, but the different proposals had never been generally accepted and the chemical literature at the time was confusing to say the least. To a man like Berzelius, with his demands for a logical and systematic presentation of chemical facts, this disorder must have been intensely irritating and he resolved to do something about it. He may not have been too fond of Latin as a schoolboy but he realized that it was the only truly international language. Consequently, when he introduced his chemical symbols in 1811 he used the first letter in the Latin name of an element, for instance C for carbon (*carbo*), while for copper, Cu (*cuprum*), he had to add one more letter. To designate the proportions of atoms making up a molecule he used index numbers, for instance H_2O for water. Berzelius' system of chemical symbols was rapidly accepted, perhaps because of the authority and general respect that he enjoyed internationally. In any case, this common chemical language was a prerequisite of the remarkable development that followed.

Gases and the Concept of the Molecule

The study of gases and their properties led to important chemical discoveries in the second half of the 18th century. This line of investigation continued during the following century and would eventually result in a fundamentally new interpretation of such concepts as atoms and molecules. It is interesting and perhaps encouraging to note that some of these important scientific developments took place during the upheavals of the French Revolution and the following long and bloody Napoleonic Wars.

French chemist Joseph Louis Gay-Lussac (1778–1850) (Fig. 6) was born into a bourgeois family with a solid position both socially and economically, but the young boy's world was brutally shattered by the Revolution when his father was arrested during the years of the Terror. Nevertheless, he managed to attend the newly opened École Polytechnique, followed by the national civil engineering school. Upon graduation in 1800 he was taken in hand by the famous chemist Claude Louis Berthollet who had accompanied Napoleon on his Egyptian expedition and was a scientist that the emperor-to-be had a high regard for. Thus Gay-Lussac received a thorough chemical training and at the same time acquired an important patron who did much for his professional advancement.

Gay-Lussac's first major scientific achievement came in 1802 when he investigated the expansion of gases as a function of temperature. He worked systematically and with great precision, using a variety of gases and repeating each experiment a number of times. Gay-Lussac found that the same increase in temperature caused equal volumes of all gases to expand equally much, i.e. they had the same coefficient of expansion. He was not the only one to make such measurements; in fact both John Dalton and Jacques Charles studied the thermal expansion of gases but their experiments were much less exact. His most important discovery as far as gases were concerned was the law of combining volumes, i.e. when gases react chemically with each other they do this in simple proportions by volume. The experiments that led to this principle were conducted in collaboration with the famous German naturalist Friedrich Heinrich Alexander, Baron von Humboldt (1769–1859), perhaps the most renowned scientist of his time. Born in Berlin of a noble Prussian family he was truly a polymath, remembered both for his geographic discoveries (the oceanic current off the west coast of South America was named after him) and his contributions to such diverse fields of science as chemistry, physiology and mineralogy, to mention just a few. During a stay in Paris he made the acquaintance of Gay-Lussac and in spite of the difference in age as well as social position, and notwithstanding the fact that Prussia and France were at war with each other at the time, they became close friends.

Fig. 6.   Joseph Louis Gay-Lussac (1778–1850).
Courtesy of the Royal Swedish Academy of Sciences, Stockholm.

Humboldt and Gay-Lussac used mixtures of oxygen and hydrogen, which were brought to react with each other in an explosion set off by an electric spark. The reaction that ensued always resulted in two volumes of hydrogen combining with one volume of oxygen to form two volumes of water vapour. These discoveries were made in 1805 and announced in a paper read to a scientific society the same year. However, it was not until 1808 that Gay-Lussac published a full account of his general law of the combining volumes of gases. It would turn out to have the greatest implications for our understanding of the relationship between atoms and molecules.

***

Italy had remained a centre of European culture and science regardless of the political confusion that had for many centuries divided it into numerous independent small states which seemed to be constantly in conflict with each other. In the 18th and 19th centuries the main point in science can perhaps be said to have moved north of the Alps but names such as Alessandro Volta (1745–1827) and Amadeo Avogadro (1776–1856) (Fig. 7) demonstrate that nonetheless important scientific research continued to be done in Italy.

Amadeo Avogadro was the son of Count Filippo Avogadro, a high civil servant and senator of Piedmont in Northern Italy. Coming from a family of lawyers Amadeo studied jurisprudence and to begin with had a career as a civil servant. However, in 1800 his scientific interests prevailed over law and he began to study mathematics and physics. In 1820 he was appointed professor of mathematical physics in Turin where he remained until his death. Unlike his famous contemporaries Berzelius and Gay-Lussac, he worked alone and never managed to form a school of young scientists around himself, while for instance Berzelius had a number of admiring pupils who ended up as professors at universities all over Europe, which very much contributed to his international prestige. It would take a long time before the scientific community realized the importance of Avogadro's work.

He is known above all for his hypothesis of 1811 that defines the difference between atoms and molecules. It takes Gay-Lussac's law of combining volumes of gases as its starting point, but Avogadro goes one important step further. He proposes that "the number of integral molecules in any gas is always the same for equal volumes, i.e. it is always proportional to the volume". From this we may infer that the ratios of the densities of two gases (at the same temperature and pressure) are the same as the relative weights of the two molecules since equal volumes of gas contain equal numbers of particles (molecules).

In a second memoir published in 1814 Avogadro advanced what might be called his second hypothesis when he proposed that compound molecules of

Fig. 7.   Amadeo Avogadro (1776–1856).
Picture from the Internet.

simple gases must be composed of two or more atoms. For instance, molecules of hydrogen, oxygen and nitrogen are made up of two atoms each. A molecule of water is composed of half a molecule of oxygen (O) and one molecule of hydrogen (H_2) to give the molecule H_2O. He went on to give the correct formulas for several compounds containing carbon and/or sulfur, for instance carbon dioxide, carbon disulfide, sulfur dioxide and hydrogen sulfide. Perhaps because of the relative isolation in which Avogadro worked in Turin, it took a long time for the scientific community to recognize that the molecular weight of gases was proportional to their density and that this could be used to determine the atomic weight. It would in fact not be until the famous scientific congress in Karlsruhe in 1860 that this was generally accepted, thanks to the efforts of his countryman Stanislao Cannizzaro.

Important Results of a Congress

In 1860 the confusion about such important concepts in chemistry as atoms and molecules and their relation to each other had become so troublesome that something obviously had to be done about it. It was decided to organize a great congress in the German town of Karlsruhe with all the leading minds of chemistry participating. This ambition was certainly realized in the sense that the congress was attended by a great number of younger chemists, among whom we can find many of the leading names in the chemistry of the future, for instance August Kekulé, Carl Weltzien and Charles Adolphe Wurtz who had in fact taken the initiative to organize the meeting. Other participants that would make remarkable contributions at the congress or whose own scientific thinking would be strongly influenced by it were the Italian Stanislao Cannizzaro, the German Lothar Meyer and the Russian Dmitri Mendeleev. On the other hand, of the senior chemists who had dominated the first half of the 19th century and were still alive (Dalton, Berzelius and Gay-Lussac had all passed on), only French organic chemist Jean Baptiste Dumas was present when the congress opened on 3 September 1860. Another disappointment to the organizers was that the meeting refused to take any definitive decision on the issues discussed. Dumas, for instance, was a supporter of the idea that concepts like atoms and molecules had a different meaning in organic than in inorganic chemistry, a theory that was in fact the subject of the lengthy talk he gave at the congress. Cannizzaro hotly denied this but a majority of the audience felt that it was after all not a political issue which could be resolved by taking a vote. In the end everyone had to make up his own mind and the question was left at that. Nevertheless, in spite of this indecision, the congress of Karlsruhe would turn out to be one of the most important chemical meetings ever, in terms of its influence on the scientific thinking of the participants.

Stanislao Cannizzaro (1826–1910) (Fig. 8) came from a well-known Sicilian family who by and large supported the rule of the Bourbon kings of Naples. However, there were exceptions to their royalist loyalty. Three of Stanislao's maternal uncles were killed while fighting in the campaigns of Garibaldi and Stanislao himself as a young man took part in the rebellion against the Bourbons. He had entered the medical school of the University of Palermo and being attracted to scientific research he felt the need for a better training in chemistry. The leading Italian chemist at the time was Raffaele Piria at the University of Pisa and here Cannizzaro got his basic chemical training. Having returned to Palermo in 1847 he became involved in the rebellion and when the Bourbons eventually triumphed he had to flee to France where he continued his studies in Paris. After two years there he was able to return to Italy and in 1855 he was appointed professor of chemistry at the University of Genoa. Here he devoted much time and thought to the course in theoretical chemistry which he gave to his students and also summarized in a famous letter to his friend and colleague Sebastiano de Luca in Pisa. This description of his course (often referred to as *Sunto*) was published in a chemical journal in Pisa and also appeared in print as a pamphlet in 1859. At the time it was rather neglected by the scientific community but a year later at the congress in Karlsruhe it would bring Cannizzaro lasting fame as a pioneer of modern chemical thinking. While this was going on, Garibaldi's Sicilian revolt had been successful and Cannizzaro could return to Palermo as professor of inorganic and organic chemistry. After the unification of Italy he moved to Rome in 1871 where he was made a senator as the climax of his distinguished career.

Like so many other young and energetic European chemists with original ideas that they wanted to advance, Cannizzaro gave a talk at the congress in Karlsruhe. It would seem that on reflection a majority of the participants felt that Cannizzaro's lecture, which he delivered on the last day of the congress, was perhaps the most important contribution made. He began by giving the historical background to the present confusing situation. Cannizzaro explained why Berzelius had rejected Avogadro's idea that the molecular structure of such gases as hydrogen and oxygen might contain two atoms each. Berzelius was completely captivated by his electrochemical theory, which required that chemical compounds always be held together by opposite electrical charges, and this was obviously impossible for Avogadro's hydrogen and oxygen molecules.

After having disposed of Berzelius' electrochemical theory as a general principle for all chemical interactions, Cannizzaro went on to explain Avogadro's hypothesis and his distinction between atoms and molecules. Moreover, he stressed how vapour densities could be used to determine molecular weights and atomic weights. At the same time he completely rejected Dumas' idea that the terms "atom" and "molecule" could have different meanings in organic and

Fig. 8.   Stanislao Cannizzaro (1826–1910).
Picture from the Internet.

inorganic chemistry. After delivering his talk Cannizzaro left at the end of the meeting but a friend of his remained behind and distributed pamphlets containing Cannizzaro's *Sunto* to the audience. This gave a number of young chemists from all over Europe the opportunity to become acquainted at leisure with the ideas of Avogadro, which had for such a long time remained strangely unappreciated. Now they took on new life and were discussed everywhere. Surely, this had to do with the fact that the time was now ripe for such new and revolutionary concepts, but it is also true that Cannizzaro was a superb communicator, something that could hardly be said for Avogadro.

Atomic Weights
and Their Relation
to Chemical Properties
of Elements

It has been said that the Karlsruhe Congress was the most important of all chemical meetings in terms of its lasting impact on many of the participants and the way they thought about their science. In any case, this was certainly true for two members of the young and enthusiastic audience, who listened to Stanislao Cannizzaro's lecture with rapt attention when he gave his revolutionary talk at the very end of the congress. They did not know each other and their backgrounds were totally different. One of them, Lothar Meyer, came from Breslau University and had been raised in the exceedingly strong research tradition of German chemistry, having for instance studied in Heidelberg under the famous Robert Bunsen. The other man, whose scientific career was profoundly influenced by listening to Cannizzaro, was Dmitri Mendeleev, who had been born in Siberia, very far from the centre of European science.

The Road from Tobolsk to St. Petersburg

Dmitri Mendeleev (Fig. 9) was born in Tobolsk, on 8 February 1834, as the thirteenth and last but one child of Ivan Mendeleev, who was headmaster and teacher of Russian literature at the local gymnasium. The latter died in 1847 after having been seriously ill for several years, leaving his wife Maria Kornileva and his large family in strained economic circumstances. Nevertheless, Dmitri was able to attend the gymnasium where he excelled in subjects like mathematics and physics, while he seems to have disliked ancient languages and theology. His mother thought very highly of her talented young son and she was determined to give him the best education she could afford. Shortly before her death in 1850 she sold all her assets in Tobolsk and set out with Dmitri on the endlessly long journey first to Moscow, where he was refused enrolment at the university on bureaucratic grounds, and then to St. Petersburg. Here Dmitri was accepted as a student at the faculty of physics and mathematics of the Main Pedagogical Institute. Fortunately the president at the Institute turned out to have been a fellow student of Dmitri's father, which probably helped things along. Shortly afterwards his mother passed away, having had the satisfaction of seeing her son embark on the scientific career that she had done her utmost and spent all her last resources to make possible. Later in life, when Mendeleev was an internationally famous scientist, he paid tribute to the decisive influence of his mother by dedicating a major scientific publication to her memory.

After his mother's death the Pedagogical Institute became the centre of young Mendeleev's life. He was only 16 years old at the time, and he seems to have devoted himself completely to his studies. Several of his teachers were well-known scientists and generally speaking the Institute must have provided an intellectually

Fig. 9.   Dmitri Mendeleev (1834–1907).
Picture from the Internet.

stimulating environment. Mendeleev was a very successful student but during his first year in St. Petersburg he became ill with symptoms suggesting an affection of the lungs, probably tuberculosis, although a definite diagnosis was not possible at the time. He was particularly interested in chemistry and his achievements were so promising that after his graduation in 1856 he was rewarded with a gold medal for excellence. His teachers therefore wanted to offer him a position at the Institute in order to facilitate his academic career. However, his state of health steadily deteriorated and his physician strongly recommended that he be sent to the South to benefit from a milder climate. Obviously his doctor did not expect him to live for more than a few months. Having spent some time as a teacher on the Crimean peninsula, Mendeleev's condition took an unexpected turn for the better and the next year he was back in St. Petersburg again. Here he was appointed a lecturer in chemistry at the university and he was generally regarded as one of its most promising young scientists.

Mendeleev's dissertation on graduating from the Pedagogical Institute had dealt with isomorphism, i.e. the fact that substances of different chemical composition could, nevertheless, show the same crystal form, a phenomenon first observed by German chemist Eilhard Mitscherlich. Mendeleev continued to be interested in the relation of the crystalline form of a substance to its other properties, for instance its specific volume. This early work showed him to be strongly influenced by French chemist Charles Gerhardt, who opposed the widely accepted view of Jacob Berzelius that the chemical bond was dependent on an electric attraction between differently charged components. Gerhardt's major contribution to the development of organic chemistry was the introduction of the idea of a homologous series. Substances with similar chemical properties, for instance the alcohols known at the time, had a significant numerical relation between their empirical formulas, suggesting that they could be seen as members of a homologous series, where each member could be derived from its nearest neighbour by the addition (or removal) of a characteristic unit. Mendeleev's early infatuation with the ideas of Gerhardt and his rejection of Berzelius' view of the chemical bond as dependent on electric attraction would have long-lasting effects on his thinking. Later in life he had difficulties accepting both the electrolytic theory of Arrhenius and the concept of the electron.

Even in the Russia of the Tsars, after the reactionary regime of Nicholas I, with the country weakened by the unlucky Crimean War, the government was still in the habit of sending unusually gifted young scientists abroad to complete their studies. When this opportunity was extended to Mendeleev in 1859 he decided to go to Heidelberg where the university boasted such names as Bunsen, Kirchhoff and Helmholtz. One might have thought that Mendeleev would have eagerly presented himself to Robert Bunsen and asked to be given the privilege to join his group and work in his famous laboratory, where young scientists from all over the world flocked in order to acquaint themselves with the latest progress in chemistry. Not so; the fastidious Mendeleev found Bunsen's laboratory to be a smelly and unattractive place to work in and preferred to turn his living quarters into a private lab where he could work according to his own ideas. It is true that he attended Bunsen's lectures but his absence from the great man's lab had been noted with disapproval and Bunsen never really forgave Mendeleev for what he considered to be a slight.

In his private laboratory Mendeleev devoted much of his time to the study of gases, particularly under conditions when their behaviour deviated from the predictions of the general Boyle–Mariotte law of ideal gases. He discovered the critical temperature, although he called it "the absolute temperature of boiling", i.e. the highest temperature where the gas could still be liquefied by the application of pressure alone. Thus, his findings can be seen as portending the later theories of liquefaction of gases based on the work of Thomas Andrews and Johannes van der Waals. These results confirmed his belief that the physical and chemical properties of compounds were intimately related to their mass. His own thinking along such lines had certainly made him well prepared when at the Karlsruhe Congress he became acquainted with the theories of Avogadro about the relationship between atoms and molecules.

From the Physiology of Blood Gases to the Mass of Atoms and Molecules

Lothar Meyer (Fig. 10) was born on 19 August 1830 in Oldenburg, a county in the north-west of Germany, where his father was a district medical officer. He was the fourth of seven siblings, but as was so often the case in those days, even in fairly prosperous middle-class families, the three elder children passed away at an early age. Lothar himself was considered a very delicate child and Dr. Meyer must have worried a great deal about the health of his eldest son. Nevertheless, Lothar seems to have enjoyed school where he was a successful student who was particularly fond of classical languages and the humanities (subjects that we know his future colleague Dmitri Mendeleev hated). When Lothar was about to enter the gymnasium he began to suffer from serious headaches, which caused his alarmed father to forbid him all intellectual exertions. Instead he was apprenticed to a gardener and told to spend as much time outdoors as possible. After having worked for a year in the garden he was in much better condition and could now return to school and attend the Oldenburg Gymnasium, where we know that he excelled in Latin and Greek, but also in the natural sciences and mathematics.

Fig. 10. Julius Lothar Meyer (1830–1895).
Picture from the Internet.

Another result of his apprenticeship as a gardener was the great joy that such work gave him in the rest of his life.

Having graduated from the gymnasium Lothar decided to follow the example of his recently deceased father and take up medical studies. He enrolled in the medical faculty of Zurich University and after completing his studies there he went to the famous laboratory of Robert Bunsen in Heidelberg to learn more chemistry. Here he came in close contact with such future leaders in German chemistry as August Kekulé and Adolf von Baeyer. He also became proficient in recent methods for the analysis of gases. This stood him in good stead when later he became interested in the binding of oxygen and carbon monoxide in the blood, which was the subject of his Ph.D. thesis at the University of Breslau in 1858. He was appointed lecturer in physiological chemistry at the university and in this position his teaching load was indeed heavy, not only including general biochemistry but also different analytical methods as well as refresher courses in organic and inorganic chemistry. This gave him little enough time for his own research but, on the other hand, it certainly required him to keep up with the development of chemistry at a time when this science was undergoing something of a revolution. In fact, Lothar Meyer's life at the time is reminiscent of the period that Stanislao Cannizzaro spent in Genoa preparing his famous *Sunto*, which formed the basis for his seminal lecture in Karlsruhe.

Thus, it is fair to say that both Meyer and Mendeleev came to the Karlsruhe Congress with an open mind, prepared to seriously consider the ideas of Avogadro, which had lain dormant for such a long time but were now about to be revived through the efforts of Cannizzaro. Many years later Meyer wrote down his recollections of those fateful days in Karlsruhe. He paid particular tribute to Cannizzaro and remembered that he had read the pamphlet containing the *Sunto* during the journey home from the congress. It was then that the scales had fallen from his eyes, his doubts had disappeared and he became convinced of these new concepts of atoms and molecules and their weights. His first major contribution to theoretical chemistry, *Die modernen Theorien der Chemie* (Modern Theories of Chemistry), appeared in 1864 and was strongly influenced by Avogadro's theories as interpreted by Cannizzaro. In the following years his thinking revolved increasingly around the challenge of trying to relate the properties of elements to the ordering of them according to increasing atomic weights. Similar attempts had been made previously by, for instance, the well-known chemist and hygienist Max von Pettenkofer but it was only after the confusion about atoms and molecules had been straightened out that real progress was possible.

In his *Modern Theories* of 1864 Meyer appended a table of 28 elements, arranged by atomic weights, demonstrating how the chemical properties of the elements changed with increasing atomic weight. In a similar table designed in 1868 and intended for the second edition of his *Modern Theories*, which did not appear in print until 1872, Meyer increased the number of elements to 52, giving a total of 15 groups, where the elements making up each group have similar chemical and physical properties. Thus, there can be no doubt that Meyer independently discovered the central principles of the periodic law of elements. In the meantime his academic career brought him to Karlsruhe in 1868 as a professor of chemistry. He stayed there until 1876 when he was called to Tübingen where he remained until his death in 1895. He married Johanna Volkmann, who not only gave him four children but also became one of his main scientific collaborators.

Competing for Recognition

While he was contemplating the principles that would eventually underlie the periodic law of elements, Meyer encountered a very common phenomenon in research: someone else was thinking along the same lines.

Dmitri Mendeleev on his return to Russia in 1861 became responsible for the teaching of organic chemistry in St. Petersburg. He immediately began to write a textbook of this subject in which he incorporated Avogadro's theories and also showed himself to be a follower of Gerhardt. In 1864 he became professor of chemistry at the Technological Institute and in 1867 he was called to the chair in chemistry at St. Petersburg University, where he remained until 1890. For his textbook in organic chemistry he was rewarded with a prize from the St. Petersburg Academy of Sciences and he used the money to marry Feozva Nikitichna Lesheva. By her he had a son and a daughter but the marriage was dissolved in 1882 after a long period of separation.

As soon as Mendeleev had been appointed professor of chemistry at the university he began to write a comprehensive textbook of the subject, *Osnovy khimii* (Principles of Chemistry), in which he also introduced the concept of the periodic law of elements. Thus, both for him and Lothar Meyer it was the writing of a textbook in chemistry that gave the impetus to present these new and revolutionary ideas. In Mendeleev's own words: "Elements placed according to the value of their atomic weights present a clear periodicity of properties."

In March 1869 at a meeting of the Russian Chemical Society, a paper by Mendeleev with the title "Relation of the Properties to the Atomic Weights of the Elements" was read to the audience by a friend of Mendeleev's. The author himself was away on a trip to inspect the cheese-making procedures employed in the countryside. This is, by the way, a typical example of his widespread interest in the economy, agriculture and industry of his beloved Russia.

In this first presentation of the periodic law the principal basis of the law was mainly the regular increment in atomic weight between neighbouring elements (Fig. 11). He then discovered three gaps in the continuous progression of atomic weights. There was one gap between hydrogen and lithium, another between fluorine and sodium, and one more between chlorine and potassium. To fill these gaps he predicted the existence of three yet undiscovered elements (helium, neon and argon). He also foresaw the discovery of elements with chemical properties similar to boron, aluminium and silicon, which he designated as eka-boron, eka-aluminium and eka-silicon respectively (*"eka"* being the Sanskrit numeral meaning "one").

Mendeleev's paper outlining his ideas about the relation between chemical properties and atomic weights were printed in the *Journal of the Russian Chemical Society* and soon was also available in German translation. When Lothar Meyer became aware of the line of thought that Mendeleev was pursuing he bestirred himself to collect his so far somewhat fragmentary work on the subject and presented his results in 1870 in the prestigious journal *Liebigs Annalen der Chemie*. We will return to this publication in the following. Mendeleev, being alerted to how far Meyer had come in his work, was induced to submit in 1870 yet another paper to the Russian Chemical Society with the title "On the Natural System of the Elements and How It May Be Used to Ascertain the Properties of Yet Undiscovered Elements". Here he introduced as an additional principle of the periodic law the valence of the elements and the compounds they might form part of. Thus he lays down: "The highest combination of an element with hydrogen or oxygen, and consequently with any equivalent element, is determined by its atomic weight and is a periodic function of that."

Consequently, he could foretell the discovery of the previously mentioned eka-boron, eka-aluminium and eka-silicon and also predict their chemical properties. Mendeleev summarized his discoveries in 1871 when he wrote an article in *Liebigs Annalen der Chemie* with the title "Die periodische Gesetzmässigkeit der chemischen Elemente". Here for the first time he definitively spelled out the concept of a periodic law/periodic system.

The new and seminal ideas published by Mendeleev and independently by Meyer did not cause the stir in the scientific community that one might have expected, at least not to begin with. However, when the elements eka-boron, eka-aluminium and eka-silicon foretold by Mendeleev were actually found and their chemical properties turned out to be in accordance with his predictions, the periodic law became increasingly accepted by chemists all over the world.

In 1875 French inorganic chemist Paul Émile Lecoq de Boisbaudran, without knowing anything about Mendeleev's work and his predictions, discovered a new element by spectroscopic analysis, which he named gallium. When Mendeleev

	Group I R₂O	Group II RO	Group III R₂O₃	Group IV RH₄ RO₂	Group V RH₃ R₂O₅	Group VI RH₂ RO₃	Group VII RH R₂O₇	Group VIII RO₄
1	H = 1							
2	Li = 7	Be = 9'4	B = 11	C = 12	N = 14	O = 16	F = 19	
3	Na = 23	Mg = 24	Al = 27'3	Si = 28	P = 31	S = 32	Cl = 35'5	
4	K = 39	Ca = 40	— = 44	Ti = 48	V = 51	Cr = 52	Mn = 55	Fe = 56, Co = 59, Ni = 59
5	Cu = 63	Zn = 65	— = 68	— = 72	As = 75	Se = 78	Br = 80	
6	Rb = 85	Sr = 87	?Yt = 88	Zr = 90	Nb = 94	Mo = 96	— = 100	Ru = 104, Rh = 104, Pd = 106
7	Ag = 108	Cd = 112	In = 113	Sn = 118	Sb = 122	Te = 125	I = 127	
8	Cs = 133	Ba = 137	?Di = 138	?Ce = 140	—	—	—	
9								
10	—	—	?Er = 178	?La = 180	Ta = 182	W = 184	—	Os = 195, Ir = 197, Pt = 198
11	Au = 199	Hg = 200	Tl = 204	Pb = 207	Bi = 208			
12	—	—	—	Th = 231	—	U = 240	—	

Fig. 11.   Mendeleev's original version of the periodic system.

claimed that gallium was really the eka-aluminium he had previously predicted, Lecoq de Boisbaudran first objected to Mendeleev's conclusion on the grounds that gallium's specific weight was lower than that calculated for eka-aluminium. However, when Mendeleev insisted, Lecoq made a second determination of the specific weight of his new metal, using a more extensively purified preparation, and this time the value determined agreed with that of Mendeleev's eka-aluminium.

The discovery of gallium and its identity with eka-aluminium did much to make Mendeleev's work more widely known in the scientific world, but this was just the beginning of a triumphal procession. Among the many scientists who to begin with refused to believe in Mendeleev's bold theorizing was Swedish chemist Lars Nilson, who thought strictly in empirical categories and saw experimental research as the only way to success. It must have come as a great surprise to him when the new metal which he had detected by spectroscopy in 1879 turned out to be identical with Mendeleev's eka-boron. In accordance with the patriotic inclinations of the time he called it scandium. After all, Lecoq had named his new element gallium! Perhaps the most convincing evidence for the periodic law came from the discovery of germanium in 1886 by German chemist Clemens Winkler. The very extensive information about the properties of this metal that he obtained undoubtedly proved it to be identical with Mendeleev's eka-silicon.

The finding and characterization of these three metals and their conclusive identification as the predicted eka-elements of Mendeleev was of decisive importance for the eventual acceptance by the scientific community of the periodicity of elements originally suggested by Mendeleev on purely theoretical grounds. It was with good reason that the fifth edition of his book *Osnovy khimii* that appeared in 1889 contained pictures of Lecoq de Boisbaudran, Nilson and Winkler, who also appeared on the wall of Mendeleev's study with the common caption "Upholders of the Periodic Law".

Furthermore, the periodic law was not consistent with the then generally accepted atomic weights for a number of elements, for instance beryllium. Bohuslav Brauner, a chemist at Prague University, had been convinced of the importance of the periodic law when Lecoq de Boisbaudran identified gallium as eka-aluminium and he resolved to devote his scientific career to "the experimental examination of the problems connected with Mendeleev's system". He started by re-examining the atomic weight of beryllium, at the time taken to be approximately 13. This did not agree with the position of beryllium suggested by Mendeleev in his periodic table, which instead indicated an atomic weight of about 9 for this element. It would then occupy a place at the head of group 2, which in addition contained elements such as magnesium, calcium, strontium and barium. Brauner's meticulous analysis showed that beryllium had indeed the atomic weight predicted by Mendeleev. Brauner also took on the complex group of elements called rare earths and he could show that these occupied a position in the periodic table between number 57 (lanthanum) and number 72 (hafnium). These two elements were isolated in a pure form first in 1923.

When the periodic law was first suggested, only cerium and terbium in this group of metals had been characterized as distinct elements. The correct placing of the rare earth elements in the periodic system proved to be very problematic, particularly since around 1880 a number of new elements were discovered in this group by spectroscopy. Among them were ytterbium, holmium, thulium and erbium. A substance believed to be didymium proved to contain the elements samarium and gadolinium, while didymium itself consisted of two elements, neodymium and praseodymium. In fact, it would not be until 1945 that all 14 elements of this group were discovered and characterized. Brauner's painstaking analyses established the extensive similarity between the rare earth metals, which of course explains the great difficulties that their separation and characterization had presented. He therefore concluded that the rare earth elements must constitute a particular inter-periodic group, usually shown as a separate row under the rest of the periodic system. His results represented an important experimental support of the periodic law and as a token of appreciation his portrait occupied a place of honour on the wall of Mendeleev's study together with the pictures of Lecoq de Boisbaudran, Nilson and Winkler. In 1871 Mendeleev had essentially completed his own work on the periodic law/periodic system.

As we shall see later Mendeleev closely followed the development in other laboratories and frequently commented on it. Of particular interest to him was of course what went on in the laboratory of Lothar Meyer and on certain occasions there were clashes of opinion in terms of which of them could claim priority to the opening of this new field of theoretical chemistry. On the whole it seems fair to say that Mendeleev had the leading edge here. He showed more imagination and scientific fantasy than Meyer and was more daring in his conclusions and predictions. However, there were certain aspects of the periodic law where Meyer was clearly the pioneer. In the paper that he published in *Liebigs Annalen der Chemie* (1870) he showed a graphical representation of the atomic volume (atomic weight divided by specific weight) as a function of the atomic weight of the solid elements. It appears that elements with similar chemical and physical properties fall on similar parts of the curve. For instance, the alkali metals (Li, Na, K, Rb and Cs) make up the peaks of the curve, while the earth metals (Be, Mg, Ca, Sr and Ba) are found on the descending, and the halogens (F, Cl, Br and J) on the ascending parts of the curve. Based on these characteristics of the curve Meyer was able to assign the correct positions in the periodic table to Au, Hg, Tl and Pb, which Mendeleev had previously misplaced. There can be no doubt that these results were of importance for Mendeleev's famous paper of 1870 "On the Natural System of the Elements" (see earlier).

It is hardly surprising that Mendeleev and Meyer should now and then get into conflict with each other over their relative contributions to the periodic law. The first time this happened was in 1879 and the dispute was actually caused by French chemist Charles Adolphe Wurtz whose book *La Théorie atomique* (Atomic Theory) had just been translated into German. In this translation the German publisher had inserted references to Lothar Meyer's work, but the author, Wurtz, objected to this, in his opinion, unwarranted emphasis of Meyer's contributions to the law. Instead he indicated Mendeleev as being the first to have come up with the idea to order the elements according to increasing atomic weights, even if Meyer had contributed important details to the law. These opinions of Wurtz's provoked Meyer to a response a few weeks later. He freely admitted Mendeleev's priority, particularly in terms of his daring predictions, but at the same time pointed out that Mendeleev's original tables had contained discontinuous rows of elements, as well as elements with incorrectly assigned atomic weights. In contrast to this, as Meyer pointed out, in his own tables the elements appeared in a single continuous row with correct atomic weights. These marked improvements he attributed in part to his analyses of the atomic volumes as a function of the atomic weights (see earlier).

Meyer's fairly innocent rebuttal seems to have hit Mendeleev on the raw and he replied in no uncertain terms that in his opinion he was the true creator of the periodic law and that he was in no way beholden to Mr. Lothar Meyer as far as the periodicity of the elements was concerned. In fact, based on a perusal of his earlier publications he came to the conclusion that Meyer never had entertained the idea of a periodic law until he read Mendeleev's papers and even after having done that, had not been able to contribute anything of importance. In Mendeleev's opinion the claim of Lothar Meyer must be regarded as a delusion. This rather ungenerous outburst of Mendeleev's can hardly be accepted as a fair assessment of Meyer's contribution to the periodic law. Instead there is every reason to believe that Mendeleev and Meyer, after having attended the important Karlsruhe Congress and having been exposed to Cannizzaro's seminal lecture, had independently of each other followed up similar ideas about the chemical and physical properties of elements as a function of their atomic weights. In any case, in 1882 the Royal Society of London seems to have reached a similar conclusion when they awarded the gold Davy Medal, one of their highest distinctions, to Mendeleev and Meyer jointly.

Mendeleev's irate comments on Meyer's claim to credit for at least some part of the periodic law would seem to have been the result of a passing mood. In years to come their relations became rather friendly and in 1887 they both attended a meeting of the British Chemical Association in Manchester. Lothar Meyer seems to have taken a more detached view of the conflict. In 1880 he wrote the following on the subject: "It is hard to be objective when somebody unexpectedly takes up your own most cherished ideas and theories. When I encountered Mr. Mendeleev's first publication in 1869 I wanted to call out to him: 'Noli turbare circulos meos!'" (The line translates as "Do not disturb my circles!", words supposedly uttered by Archimedes to the enemy soldiers who brutally murdered him, when after having taken Syracuse they found the Greek mathematician drawing geometrical figures in the sand in his garden.)

Unexpected Support for the Periodic Law

The discovery of the inert gases is a good example of how new chemical information (these substances were completely unknown at the time that the periodic law was conceived) to begin with seemed to contradict the law but in the end turned out to support it.

Henry Cavendish (1731–1810) was one of those deeply original amateur scientists who have made such important contributions to British science. He came from an influential family (his uncle was the 3rd Duke of Devonshire) and spent four years at Peterhouse College in Cambridge, which he left without taking a degree. To begin with he lived on a moderate allowance from his father, but eventually a legacy from an aunt made him immensely rich so that he was said to have been the richest of philosophers and the most philosophical of the rich. His habits, nevertheless, continued to be very frugal, he never married and could thus completely concentrate on his research where he became a pioneer in the study of gases. In this field considerable progress was made during the 18th century, not least through the efforts of Cavendish.

During his studies of the gases that make up the atmosphere of the earth he attempted to completely oxidize the atmospheric nitrogen in the presence of an excess of oxygen by sparking the air with electricity from a static machine. Regardless of how long he continued the sparking experiment there was always a small residue of gas, corresponding to about 1/120 of the volume of atmospheric nitrogen, which seemed to be resistant to oxidation. What he had there was in all probability an inert gas, but it would take another century before the first of these gases, argon, was actually discovered.

John William Strutt (1842–1919), 3rd Baron Rayleigh and a peer of the realm by inheritance, belonged like Cavendish to the upper crust of the British nobility, perhaps not a position in society that would normally have predestined him for a scientific career. Nevertheless, he became one of the dominating figures in 19th century physics, on the same level as Helmholtz, Gibbs and Kelvin. Rayleigh was active in almost all major fields of physics and held a professorship in this subject at Cambridge from 1879 to 1884. However, in 1868 he set up a research laboratory at the family seat in Terling Place and when he succeeded to the title after his father he retired from his chair at Cambridge. He settled for good in Terling and the manor house laboratory there remained his scientific headquarters until his death.

His research included among many other things measurements of the density of gases, which led to the discovery of argon. This came about in the following way. When Rayleigh made high precision measurements of the density of nitrogen in order to obtain better values of the atomic weight of this element, he found that the density of nitrogen prepared from ammonia was slightly less than the density of atmospheric nitrogen. He then recalled the experiments of Cavendish indicating the presence of something in the air that unlike atmospheric nitrogen could not be oxidized by electric sparking in the presence of oxygen. He repeated Cavendish's experiment using a better source of electricity for the sparking, but he nevertheless had difficulties producing enough of the residual gas for a thorough analysis of its properties. It is now that we encounter his colleague and sometime competitor, Sir William Ramsay.

William Ramsay (1852–1916) (Fig. 12) was born in Glasgow into a family with a tradition of working as chemist-dyers. He was raised in the Calvinist tradition and when he matriculated at the University of Glasgow he took courses in the classics as preparation for theological studies and a career in the ministry. However, his interest in science eventually made him take up physics and chemistry instead. In 1871 he spent a year at the University of Tübingen where he studied organic chemistry under Rudolf Fittig. Having held different academic positions in Glasgow and Bristol he was in 1887 appointed to the chair in chemistry at University College London and here he remained until his retirement in 1912. During this time he improved the level of both teaching and research at the College and his own work on the inert gases made him internationally famous.

He had been fascinated by Lord Rayleigh's presentation to the Royal Society in April 1891 concerning the discrepancy between atmospheric and chemical nitrogen (see earlier). Ramsay claims to have asked Rayleigh at the time if he had any objections against Ramsay trying his hand at discovering the cause of the discrepancy in density between the two types of nitrogen. Having been given permission to go ahead, or so he claims, he did just that and spent the summer looking for a heavy gas in atmospheric nitrogen. He first removed oxygen from the air by electric sparking and then got rid of the nitrogen by combining it with hot magnesium to form magnesium nitride. What remained was a gas with a density of 19 to 20 (relative to hydrogen which was taken to have a density of 1) and an astonishing inertness to chemical reactions. Lord Rayleigh and Ramsay by mutual agreement wrote a joint article with the title "Argon, a New Constituent

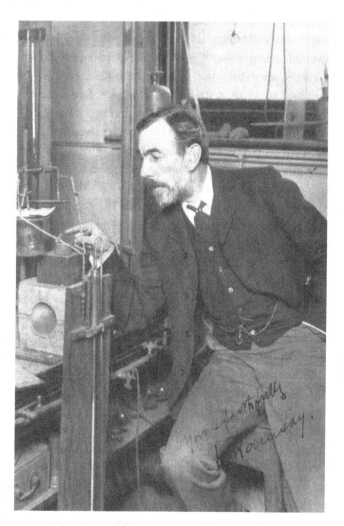

Fig. 12.   William Ramsay (1852–1916).
Picture from the Internet.

of the Atmosphere", which appeared in *Philosophical Transactions of the Royal Society* (1895). They found argon to be monatomic and to have an atomic weight of approximately 40.

In 1868 astronomers Pierre Jules Janssen and Joseph Lockyer by spectroscopic analysis of a solar protuberance had discovered a characteristic yellow line indicating a previously unknown element in the sun, later called helium after the place of its discovery. Ramsay in 1895 found terrestrial helium in a mineral named after Swedish chemist Per Cleve and this discovery was independently confirmed in Cleve's laboratory the same year. Like argon, helium was found to be monatomic and to show a marked chemical inertness. Thus, there were now two inert gases, argon and helium; the question was if they could be accommodated in the periodic system. This problem was particularly acute in the case of argon, which based on its atomic weight (39.95) had to be squeezed in between potassium (39.10) and calcium (40.08) where it seemed to constitute a breach in the general regularity imposed by the periodic law. Ramsay pointed out in 1897 that in all likelihood there must exist a third inert gas with properties intermediate between those of helium and argon. The next year Ramsay and his collaborator Morris Travers examined the residue left after the evaporation of liquid air. Having removed all remaining nitrogen and oxygen they found the residue to contain a chemically inert gas with a characteristic spectrum of green and yellow lines, which they named krypton. Using a similar procedure starting with a liquid argon residue they observed spectroscopic evidence of yet another inert gas, neon. When they had obtained sufficient quantities of these gases for chemical characterization, they observed in the krypton fraction tiny amounts of an even less volatile gas with a spectrum of blue lines and they named the new gas xenon. Thus, Ramsay and his collaborators had in five years isolated and characterized five chemically inert, monatomic gases: argon, helium, krypton, neon and xenon. Since then argon has been given the atomic number of 18 and placed between chlorine and potassium. This family of inert gases (now supplemented by the radioactive element radon) forms a group of its own, at the time designated as group 0, now usually called group 18. It can be said to close a gap in the periodic system, a gap that neither Mendeleev nor Meyer had been aware of. At the same time it further strengthened the experimental evidence for the periodic law.

Ramsay was indeed recognized by the scientific world. The Royal Society had elected him a fellow in 1888 and in 1895 they awarded him the Davy Medal. He was knighted in 1902 and received medals and honorary degrees from all over the world. In 1904 he was awarded the Nobel Prize for Chemistry, while at the same time Lord Rayleigh received the Nobel Prize for Physics. In what follows we will consider the implications of these Nobel Prizes in more detail.

Straightening Out Some Irregularities

At the previous turn of the century the periodic law was generally recognized as a scientific principle of steadily increasing importance. On the other hand, there were some anomalies in the ordering of the elements according to atomic weight. For instance, in Fig. 1 argon (atomic weight 39.95) comes before potassium (39.10), cobalt (58.93) comes before nickel (58.69), and tellurium (127.6) before iodine (126.9), when they are ordered according to their atomic numbers in the periodic system. This has to do with the presence of isotopes, first observed by Frederick Soddy in 1912 when he studied the radioactive disintegration of different elements. Incidentally, such changes of the elements were completely against the chemical principles upheld by Mendeleev and he had great difficulties accepting this. The radioactive disintegration of for instance uranium gives lead as the end product and this is true also for the disintegration of thorium and actinium. However, in the case of uranium it gives lead with an atomic weight of 206, while thorium gives lead 208 and actinium lead 207, respectively. Soddy coined the term "isotopes" to designate this phenomenon where the same element could show different atomic weights. The occurrence of isotopes was explained when James Chadwick in 1932 discovered the neutron, which has a mass very close to that of the proton but no electric charge. In the case of the lead isotopes discussed earlier the atomic nucleus contains the same number of protons, which gives the isotopes the same atomic number and identical chemical properties. However, the neutron number differs and this explains the different atomic weights of the isotopes. A systematic investigation has shown that almost all elements so far discovered have isotopes.

To summarize, the complicated modern atomic theory with its hundreds of different nuclear particles and its wealth of nuclear reactions is nevertheless built principally on the planetary model formulated in 1911 by Ernest Rutherford (1871–1937) (Fig. 13) and the theory proposed by Danish physicist Niels Bohr (1885–1962) (Fig. 14) in 1913 as well as in later communications. Rutherford estimated that practically all the mass of the atom was contained in its nucleus, whose diameter was only an infinitesimal part of that of the atom itself. Outside the nucleus, shells of electrons are found with a total negative charge equal to the positive charge of the nucleus. Bohr predicted that the electron shells represented discrete energy levels and that the electrons in the outer shell mainly determined the chemical properties of the element. The periodic recurrence of these properties characteristic of the periodic system is caused by the repetition of the structure of the outer electron shell.

Fig. 13.   Ernest Rutherford (1871–1937).
From Les Prix Nobel en 1908.

Fig. 14.   Niels Bohr (1885–1962).
Picture from the Internet.

Life After the Periodic Law

The scientific world may have been slow in accepting the periodic law and realizing fully its wide implications for almost every aspect of chemistry. Nevertheless, there can be no doubt that the discovery of the law did much to facilitate the scientific careers of both Lothar Meyer and Dmitri Mendeleev. In 1876 Meyer received the very attractive offer to succeed Rudolf Fittig as professor of chemistry at the prestigious University of Tübingen. The beautiful situation on the river Neckar appealed to him and here he would be given ample opportunities to assemble a large group of collaborators and students. Meyer remained in Tübingen until his sudden death from a stroke in 1895 and was obviously highly appreciated both by his students and scientific collaborators, as well as his academic colleagues. During this period his research interests came to comprise almost all branches of chemistry and his experimental work included also crafts such as glassblowing to make his own chemical apparatus. He even found time to be the supervisor of no less than 60 dissertations and to work strenuously in his beloved garden. During the last year of his life he was elected president of the university and he seems to have been in excellent health until the day that he suffered the unexpected cerebral insult that ended his life.

His book *Modern Theories of Chemistry* appeared in a great number of editions and the volume grew in size from 147 pages in 1861 to over 600 in the 1884 edition, a rate of growth that caused its author to lament that the first slender edition had been the best one. Nevertheless, there is every reason to believe that the book played an important role in the dissemination of the ideas of the periodic law. With his unassuming personality and subdued ambitions Lothar Meyer stands out as a rather attractive character in the history of science.

Lothar Meyer may indeed have been active in almost every chemical field but of his great colleague and sometime competitor Dmitri Mendeleev it is fair to say that he came to take an interest in all aspects of culture and education, as well as the industry and economic growth of his beloved Russia. Having completed his experimental work directly connected with the periodic law he turned in 1872 to an investigation of the compressibility of gases that aimed at an examination of the laws of Boyle–Mariotte and Gay-Lussac and the theories of Avogadro. The ultimate goal was to arrive at a general law for gases, but this ambitious project in the end proved to be beyond his resources. In fact, it was not realized until Johannes van der Waals presented his equation in 1880. An upshot of Mendeleev's interest in gases was his project to prove the existence of the hypothetical "universal ether" that was supposed to be the medium for the propagation of light.

In spite of the widespread belief in this "ether" it had never been demonstrated experimentally, nor did Mendeleev himself have any success in his efforts to isolate and characterize it. Even so, he had already thought of a place for it in his periodic system, somewhere ahead of helium in group 18 with an extremely low atomic weight. It was in connection with the eventual understanding of the lack of existence of ether that Einstein formulated his famous dictum "Subtle is the Lord, but malicious He is not."

Many of Mendeleev's chemical peers, for instance his successor at the Technological Institute of St. Petersburg, Friedrich Beilstein, took a dim view of his work on gases and of his capacity as a chemist in general. He was seen as being much too fond of unfounded speculations resulting in half-baked theories. When his name was put up for election to the Academy of Sciences, founded by Tsar Peter the Great in 1725, more conventional chemical candidates like Beilstein himself repeatedly defeated him. On the other hand, his international reputation as a renewer of theoretical chemistry increased all the time. When he, at the invitation of the Chemical Society of London, gave a Faraday lecture on the twentieth anniversary of his discovery, he could conclude that the law was now generally considered proven. In the same way, in Russian intellectual society, with the exception of certain envious academic circles, he was recognized as a leading figure of chemical science and his advice was avidly sought on a number of questions.

The autocratic regime of the tsars had become more favourable to reform under Alexander II, who actually had abolished the serfdom that had been the greatest obstacle to social progress in general in the Russian countryside and specifically to more efficient agriculture. Mendeleev was deeply concerned by these difficulties, which he sought to counteract by his own publications on better methods of agriculture. At the same time he tried to demonstrate the practice of such improvements on the country estate that he had acquired and now wanted to turn into a model of progressive agriculture. Another interest of his was the development of the Russian oil industry in Baku along the same lines that he had studied during a visit to the world fair in Philadelphia in 1876. In the same vein he tried to improve the mining of coal in the Donets Basin.

With the murder of Alexander II in 1881 and the accession to the throne of his reactionary successor Alexander III the short period of reform came to an end. The change of climate in society was felt also at the universities and in 1890 it resulted in political unrest among the students in St. Petersburg. Mendeleev, although in many ways of a liberal persuasion, can hardly be said to have been a revolutionary. Nevertheless, he took it upon himself to act as a mediator between the inflamed students and representatives of the government. When his efforts came to nothing he tended his resignation, which was accepted by the authorities who probably felt that he had sided with the rebellious students. This change in his academic status did not make much difference to him economically. The reason was that according to the agreement for the divorce from his first wife, Feozva Nikitichna, she received all of his salary from the university so that Mendeleev had to rely on such sources of income as royalties from his books and earnings from his activities as a consultant. In 1882 he married a young art student, Anna Ivanova Popova, whom he had courted since the late 1870s. By her he had two sons and two daughters and one can easily understand that Mendeleev sometimes had difficulties supporting his large family. It was therefore important for him when the government made use of his expertise in a number of fields, ranging from the production of smokeless powder to the system of weights and measures used in Russia. His reports on these different subjects were obviously appreciated by the authorities. In 1891 he was appointed privy councillor and two years later he became director of a newly created government office, the Central Bureau of Weights and Measures, a position that he held until his death.

Mendeleev had been a figure of contest his whole life. Even in his own country, where he had a number of admirers who looked up to him as one of the greatest representatives of Russian science, he was criticized and even ridiculed by his academic enemies. On the other hand, his acclaim by the international scientific community grew incessantly. The list of his publications contains more than 400 articles and books, he was awarded doctorates by such prestigious universities as Oxford and Cambridge, and in 1905 the Royal Society of London gave him the Copley Medal. However, he was never awarded the Nobel Prize although it had existed for six years when he died on 2 February 1907. Today no one would question that the periodic law represents just the kind of original and pioneering work that we associate with the Nobel Prize. Surely, this was realized even a century ago and if that was the case, why did not the Royal Swedish Academy of Sciences act on this insight?

THE ELUSIVE NOBEL PRIZE

The Birth of An Academy

The idea that a successful scientist deserves to be recognized and given some kind of reward for his discoveries must certainly go back a long time in history. When the medieval alchemist, having by some mysterious process obtained the philosopher's stone and discovered a way to make what appeared to be gold, turned to his prince and protector with this new and precious knowledge, he of course expected to be generously recompensed for his labours. During the period of the scientific revolution in the 17th and 18th centuries, academies of sciences started to appear in culturally leading countries on the continent of Europe as well as in Great Britain and even in Russia under the westernized Tsar Peter the Great. It now fell to these academies to support and encourage scientists by handing out stipends and rewards in the form of, for instance, prizes and gold medals. Sweden was certainly not situated at any of the cultural crossroads of Europe, but even in this country far removed from the great scientific centres, an academy of sciences was created that was destined to become the custodian of what are today among the most well-known of all scientific rewards — the Nobel Prizes in Physics and Chemistry.

The Swedish aspiration to become a great power at least in Northern Europe had come to a tragic end after the death of Charles XII and the catastrophic peace with Russia in 1721 where the country lost most of its Baltic empire. The transition from Carolingian absolute monarchy during the second half of the 17th century to the parliamentarism of the 18th century, the so-called Age of Liberty, represented not only a complete political about-turn, but also meant a re-orientation of the whole society. At least in Western Europe this was the Era of Enlightenment with ideas of liberty and equality that would come to a climax during the French Revolution. However, for the new men in power in Sweden the dominating problem was the generally weakened condition of the country. First of all it was a question of re-building the Swedish economy that was in ruins after the endless wars. Trade and industry, as well as husbandry, was the first priority for the politicians of the Age of Liberty and in spite of all the parliamentary infighting and conflicts great efforts were really made here. The successful merchant and mill owner became the heroes of the time.

This concentration on economic realities, husbandry, manufacturing and trade, rather than military glory and conquests, had great consequences also for the development of the sciences in Sweden. The leading politicians at the time obviously realized that in order to have inventions of industrial importance and better methods of agriculture, you also needed progress in basic sciences like physics, chemistry and botany. It is true that there were at the same time tendencies to discount the importance of basic research, for instance the establishment of the Commission for Education, which in 1750 put forward the proposition that the universities should be concerned only with strictly professional education, such as the training of physicians, while research in different scientific fields should be the responsibility of the newly established Royal Swedish Academy of Sciences. However, these ideas were never realized. Instead there was a steady integration of research and the education of students at the universities, in fact the same development that we have had to this day. It is not a coincidence that it was during the often denigrated Age of Liberty with its parliamentary parties, often subsidized by foreign powers like France and Russia, its corruptible officials and incessant political conflicts, that Swedish science for the first time made its appearance on an equal level with the scientifically leading nations of Europe.

The scientific revolution of the 17th century had one of its most important centres in England with such names as Isaac Newton, Robert Boyle and William Harvey. To boot it was here that the Royal Society of London was founded in 1660 and rapidly became a model for similar scientific academies on the continent of Europe. Naturally enough even in Sweden men with a vision wanted to take the same initiative. Physicist and engineer Mårten Triewald (1691–1747) (Fig. 15) had worked for several years in England and been deeply impressed not least by the scientific, technological and economic development in what he considered a leading country. Having returned to Sweden he attracted attention by the lectures in natural sciences that he delivered at the House of Nobility in Stockholm and which were also published as a book. Triewald got in touch with Carolus Linnaeus, who was at the time a general practitioner in the capital, as well as Baron Anders Johan von Höpken (Fig. 16), a rapidly rising star in the conservative party, which had just been victorious in the last election. What Triewald had in mind was the creation of a Swedish scientific academy on the pattern of the Royal Society, which he had been elected to a couple of years earlier. Like its English counterpart, which it was trying to emulate, the Swedish Academy of Sciences should be a private association and not an official governmental institution like the Prussian or Russian academies. Its purpose should be to work for the improvement and dissemination of proficiencies that were important for the realm, such as mathematics, science and economy, and furthermore to promote manufacturing and trade, a programme entirely in the spirit of the age. The first meeting of the Academy was in June of 1739, and two years later the influential von Höpken was able to secure a royal confirmation of its charter.

The Royal Swedish Academy of Sciences was a great success from the very beginning. The number of fellows increased continuously until in 1762 it was fixed at a maximum of 100. Many of the members represented of course the world of

Fig. 15.   Mårten Triewald (1691–1747).
Courtesy of the Royal Swedish Academy of Sciences.

Fig. 16.   Anders Johan von Höpken (1712–1789).
Courtesy of the Royal Swedish Academy of Sciences.

learning, but there were also a fair number of physicians, apothecaries, commoner officials and clergymen. The largest group, however, amounting to 19% of the fellows, was made up of members of the nobility and the great financial magnates. Thus, there was no lack of influential patrons and what is more, the crown prince Adolf Fredrik soon became the royal protector of the new academy. Even so, the Academy of Sciences never lost its character of a private society. The president of the Academy was elected only for a period of three months and this office was therefore mainly a token of honour for outstanding scientists like Linnaeus, or often enough for such Lords of the Realm as Count Carl Gustaf Tessin or Anders von Höpken. Therefore the position of permanent secretary became very important, particularly when this post was taken over by Pehr Wilhelm Wargentin (1717–1783) (Fig. 17). Originally an astronomer with a thesis about the moons of Jupiter among his merits, he was from 1749 to his death the administrative leader of the Academy, which he represented in its contacts both with governmental officials and foreign academies.

The Academy had been off to an early start with a number of activities, some of them rather expensive, and in the long run it could not manage with only private donations to sustain it. Through its intimate relations with leading conservative politicians such as Tessin and von Höpken, the Academy managed to obtain a privilege for the publication of almanacs. This proved to be an excellent source of income although it required a considerable amount of work, but on the other hand, here the Academy could always rely on Wargentin. As all other academies it published a series of papers that had been submitted to the Academy, *Proceedings of the Royal Swedish Academy of Sciences*, which proved to be of great importance to Swedish science. As one might expect the papers received were of varying quality and during his long period as secretary, Wargentin had the arduous task of separating the wheat from the chaff. The result must be said to have been satisfactory since the *Proceedings* were translated into several languages and had a considerable international circulation. The central position of the Academy in Swedish intellectual life remained unshaken during the whole Age of Liberty, and in the following sections we will deal briefly with some of its best-known scientific members.

The King of Flowers

If one were to make a public opinion survey and ask for the greatest Swedish scientist of all ages the answer would undoubtedly be Carolus Linnaeus, regardless of whether one interviewed the Swedish or international public. His fame has lasted to this day and there is still a Linnean Society active in London. No other

Fig. 17.   Pehr Wargentin (1717–1783).
Courtesy of the Royal Swedish Academy of Sciences.

Swedish scientist has come even close to the world fame of the brown-eyed, lively and short-of-stature clergyman's son from the province of Småland in southern Sweden. He so completely dominated Swedish science in the 18th century that one has difficulties catching sight of a number of important scientists who were his contemporaries.

Perhaps this is also because his achievements are so very much in accordance with the spirit of the age and its fascination with the appearance (morphology) of plants and animals, as well as the systematic classification of nature's wonders on the basis of morphological observations. This was, after all, the age of the laboriously gathered collections of more or less unique natural-history objects that had to be ordered according to some kind of principle. The artificial but logically constructed and therefore very useful sexual system of Linnaeus had a profound influence on the development of botany. Through his achievements, not to mention the untiring work of his many pupils, the interest in this science increased all over the world in an unprecedented way. No other science has been that popular also among the laymen. The fact that his system in this latter age does not fully meet the standards of a systematic analysis of plant DNA does not really detract from his contributions as inspirer and leader in the heyday of botany.

Carolus Linnaeus (1707–1778) (Fig. 18) grew up in the parish of Stenbrohult as the son of a clergyman, himself an interested amateur botanist who had a garden with a number of rare flowers. His father early on began to teach the boy the elements of botany. Obviously the vicar's teachings were not wasted on his son and when Carolus at the age of seven entered school in the nearby town of Växjö his interests were entirely directed towards the natural sciences. Like so many other great scientists-to-be in those days, he hated the soul-destroying Latin swotting of the traditional education and he never became much of a shining light at school. Luckily he had a protector and patron in the district medical officer Johan Rothman, who combined his medical activities with those of a science teacher at the grammar school, a fairly common arrangement at the time. He took the boy in hand and gave him private tuition when the poor vicar began to despair of his son's schooling. Rothman further strengthened Linnaeus' interest in botany and also pointed him in the direction of medicine, where Boerhaave in Leiden was the towering figure of the period.

At the age of 20 Linnaeus matriculated at the University of Lund in order to study medicine. Fortunately enough he boarded and lodged with one of the professors of medicine, Kilian Stobaeus, who obviously took an interest in the young man. Stobaeus, like so many other scientists at the time, had a passion for collecting all sorts of natural-history objects and curiosities and eventually he donated his collection to the university, where strangely enough he ended up as a professor of history. He gave Linnaeus free access to his collection and the

Fig. 18.   Carl von Linné (1707–1778).
Courtesy of the Royal Swedish Academy of Sciences.

time Linnaeus spent in the home of this original man of learning undoubtedly was of great importance for his scientific development. When after a year in Lund, following the advice of his old teacher Johan Rothman, Linnaeus moved to Uppsala University, he found the new surroundings considerably less attractive than Lund, at least to begin with.

The faculty of medicine at Uppsala at this time was in a state of decline and this was true also of the botanical garden originally laid out by Olof Rudbeck. Its two professors were old and tired and one of them, Rudbeck Jr., was also perpetually on leave to pursue his linguistic interests. As always, the young medical student was short of money, but Linnaeus' unfailing ability to find benevolent patrons again asserted itself. He must have had an unusually charismatic personality that drew people's interest to him. The dean and amateur botanist Olof Celsius happened to come across Linnaeus while walking in the overgrown botanical garden. He was immediately attracted to the unknown student and was impressed by his botanical knowledge. Linnaeus was invited to lodge in Celsius' home where he began to give private tuition in botany and through the mediation of the dean he also made the acquaintance of Olof Rudbeck Jr. The professor allowed Linnaeus to give public demonstrations in botany and furthermore hired him as private tutor for his sons. His sustenance thus being secured, there followed a few years of intense work and exceptional creativity when Linnaeus laid the foundation of his future position as the great systematist of natural science.

Linnaeus started with the plants and he was certainly not the first to take on this problem. Italian physician and botanist Andrea Cesalpino, who by intuitively having suggested the possibility of a blood circulation can be seen as something of a predecessor to William Harvey, had as a botanist come up with an "artificial" classification system based on the characteristic properties of the fruit. Others, like Swiss botanist Caspar Bauhin, had aimed at a "natural" system where the plants were grouped according to a general conformity of their morphology. Another proponent of a natural system with both species and genera was Englishman John Ray. There was also French professor Joseph de Tournefort who had in 1700 introduced a successful system that involved the morphology of both the corolla and the fruit.

The name of Linnaeus is of course inseparably connected with the so-called sexual system in botany, but he was certainly not the first to suggest this idea. German botanist Rudolph Camerarius had in 1694 demonstrated that the male sexual organ in animals had an equivalent in the stamen in the plants, while the female organ corresponded to the pistil. We know that Linnaeus was aware of these theories about the sexuality of plants and in 1729 he published a short treatise on the subject, *Prelude to the Nuptials of the Plants*. A year later he launched the idea of using characteristics of the stamen and the pistil as the basis of a classification of the plants. Eventually this idea would bear fruit in his famous *Systema Naturae*, but first he undertook some important journeys both within Sweden and to the Netherlands. The first of these journeys took him to Lappland, the northernmost and also the least known of all Swedish provinces. This long and arduous expedition was a great success both scientifically and in terms of a cultural inventory of what had previously been a practically unknown part of the country. The scientifically most important of his journeys was to the Netherlands in 1735 and the immediate purpose of it was really very prosaic: to become doctor of medicine at some suitable university there. In reality it took him only a few weeks at the University of Harderwijk, well known for handing out doctorates without too exacting demands. That detail having been taken care of and having paid his respects to the old Boerhaave, Linnaeus felt free to contact all the outstanding botanists in the Netherlands. At this time he had his great work *Systema Naturae* ready as a manuscript and when he showed it to the wealthy amateur botanist Johan Gronovius, the latter was so impressed that he offered to pay for the printing of it. Through the mediation of Gronovius, Linnaeus made the acquaintance of a new patron-to-be, the fabulously rich director of the Dutch East India Company, George Clifford, who had laid out an unparalleled botanic garden on his estate, Hartecamp. Clifford hired Linnaeus as a prefect for the garden and during two years of intense activity the young botanist was able to concentrate entirely on his research.

His scientific productivity during these few years in the Netherlands is almost incredible. Already at the end of 1735 he publishes his *Systema Naturae* where he orders the three natural kingdoms (the stone, plant and animal kingdoms) according to his own principles. The way he orders the kingdoms of stone and animals is perhaps not so revolutionary, but with the plant kingdom the sexual system makes its first appearance on the scientific scene and is presented with its 24 classes, based on the stamens of the flowers, and its orders according to the number of styles. An artificial system, no doubt, something that Linnaeus was well aware of, but at the same time a great step forward because of its logical construction and practical usefulness. Then follow a number of botanical treatises, which quickly establish Linnaeus as a leading botanist and a celebrity whom everyone seems anxious to meet. The most important of these books are his account of the flora of Lapland, *Flora Lapponica*, and the lavishly illustrated description of his patron Clifford's botanical garden and herbarium, *Hortus Cliffortianus*. In the spring of 1738 he bids farewell to the old and seriously ill Boerhaave, whose favourite pupil Linnaeus claims to have been, and returns to Sweden after a short stay in Paris. It is hardly an exaggeration to say that the really pioneering part of Linnaeus' scientific life's work was performed during these remarkable early years in the Netherlands.

On his return home Linnaeus was disappointed to find that his newly acquired world fame had not yet reached his backward native country. To begin with he was compelled to support himself as a general practitioner in Stockholm, but it did not take him long to acquire influential patrons there, such as the leading conservative politician Count Carl Gustaf Tessin, who remained a patron of Linnaeus' for life. At long last in 1741 the medical faculty in Uppsala was able to pension off the decrepit holder of one of the chairs in medicine and the young botanist was appointed professor of medicine. A year later Linnaeus made a deal with his colleague and later on great friend, Nils Rosén, whereby they divided up the teaching in such a way that Linnaeus took botany and the natural sciences with pharmacology and dietetics, and the rest was the responsibility of Rosén.

In the meantime Linnaeus had married the daughter of a wealthy physician and of the considerable number of children they had together the majority reached adult age, a fairly unusual achievement in the 18th century with its frightening child mortality. His wife was a very competent householder but an entirely different personality compared to her charming and loveable husband. Gruff and strict, she was far from popular with Linnaeus' many pupils who were happy to visit their beloved master and were frequent guests at his home, but always tried to keep as far away from his wife as possible. Linnaeus' relations with his students were warm and personal and there is no mistaking their enthusiasm for their highly thought of teacher and his subject. During the summer months, weather permitting, they made excursions in the surroundings of Uppsala with the great botanist in a van and his lectures at the Gustavianum drew full houses from all the faculties of the university. The run-down botanic garden was reconstructed with the help of a famous architect and soon became one of the leading institutions of its kind in Europe. In the new buildings Linnaeus and his rapidly growing family had their winter residence but in the summer they moved to one of the rural estates that the increasingly wealthy Linnaeus had been able to purchase. He frequently invited his young collaborators to visit him there for seminars and excursions, and in the evenings they seem to have amused themselves with barn dances and other country diversions.

Linnaeus really looked after the interests of his pupils as if they were part of his own family. With very few exceptions he was like a father for his disciples and tried in every way to further their careers. In return they regarded their master with almost religious veneration and did everything in order to spread his teachings all over the globe. They were indefatigable in the collection of plants and animals, which they classified according to Linnaeus' system, of which they had achieved a complete mastery. There was no one among them who can be said to have been on a level with Linnaeus himself, but then many of the most talented of them died very young on their perilous journeys. In any case, their efforts undoubtedly greatly increased our knowledge of the flora and fauna of the world. Starting at the middle of the 18th century they visited a number of countries that very few Swedes had even heard of and some of them participated in the most famous expeditions of the time.

The Advent of Chemistry in Sweden

The international fame of the young Swedish Academy of Sciences was of course based above all on Linnaeus, who represented botany and medicine. However, there were a number of outstanding members who were active in, for instance, astronomy, mathematics and physics. It is nevertheless fair to say that besides Linnaeus it was the Swedish chemists who made the Academy particularly influential in 18th and early 19th century science. In the following we will consider how this subject developed in Sweden.

Chemistry in the age of Paracelsus was a mixture of practical experimental knowledge, particularly in inorganic chemistry, and fantastic notions represented by alchemy and its belief in extracts and potions that could transform base metals like lead and mercury into gold. Not to mention Paracelsus' own theories about the three chemical principles: sulfur, mercury and salt, whose proper balance in the body determined illness and health. A pioneer of modern chemistry like Robert Boyle had contemptuously rejected such ideas, but the first Swedish chemist known to us, Urban Hiärne (1641–1724) (Fig. 19), was a fanatical follower of Paracelsus during his whole life and never tolerated any criticism of the great master.

Hiärne was born in a little village in Ingermanland (a part of present day Russia — including Saint Petersburg — which during the 17th century belonged to Sweden) where his father was a parson. He grew up under strained economic conditions in the outskirts of the Swedish realm and when his father died the enterprising boy betook himself to Narva, the provincial capital, and from there to Stockholm. Despite his impecunious condition he managed to enter the gymnasium in Strängnäs with the help of the principal, who obviously recognized the exceptional talents of the boy. He then studied medicine in Uppsala under Rudbeck and here he succeeded in becoming the centre of a merry social life in spite of his poverty. Perhaps he was able to support himself during his five years at the university by writing a steady stream of pastoral poems and idyllic short stories in a genre very popular at the time. Hiärne was always quite good at finding influential patrons, for instance Count Claes Tott, whom he accompanied on his travels in Germany in the capacity of physician-in-ordinary. In the years 1669–1674

Fig. 19.   Urban Hiärne (1641–1724).
Courtesy of the Royal Swedish Academy of Sciences.

he went on an extended trip in Europe to study medicine both on the continent and in England, where he managed to get himself elected to the Royal Society. Later he studied in Paris for three years and he devoted himself in particular to what was to become his main interest in life, chemistry. Having returned to Sweden he opened a successful practice in Stockholm and then followed a most remarkable period in his life when he became a member of the Witch Commission of the capital.

Witches and their dealings with the Devil was not a new problem in Europe. On the contrary, it had engaged both the church and the temporal powers for centuries. At the end of the 15th century a kind of textbook on combating witchcraft was actually published: *Malleus Maleficarum* (The Hammer of Witches). The Lutheran Reformation hardly improved the situation. In Luther's teachings the Devil had an even more prominent position than before. Even so, it was shocking and unexpected when at Whitsuntide 1668 stories about how witches took children for a ride to the Devil began to circulate in Dalecarlia and rapidly spread in the country like a spiritual pestilence. There had of course been such incidents before, when women had been indicted as witches and occasional death sentences had been pronounced, but this wholesale outbreak was something new and frightening.

The authorities realized that this evil would have to be nipped in the bud, and to that end special Witch Commissions were set up to inquire after witches, examine them and pronounce sentence on them. The priesthood further inflamed the already hysterical mood with their fire-and-brimstone sermons. Panic-stricken children testified before the Commissions to having been carried off to the Devil by witches (under normal conditions underage children were not allowed to give evidence in Swedish courts). The accused women were tortured until they had fully confessed their misdeeds, another deviation from normal Swedish legal usage at the time (unless it was a question of high treason or lese-majesty). After these irregular proceedings the alleged crimes of the poor women were considered fully proven and the death sentence was the only possible penalty. Witches were burnt at the stake all over the country (most of the time they were executed before being burnt), but even if many of the condemned were reprieved for different reasons it has been calculated that something like 300 human beings were executed, most of them women.

This madness went on year after year, but finally discerning members of the Commissions began to have doubts. The leading legal expert, Gustav Rosenhane, Chairman of the Commission for Northern Sweden, was one of the judges that voiced their doubts about the irregularities of the procedure. His very influential booklet, *Votum uthi thruldombs saken* (Judgement about the Witchcraft Case), was published in 1673 and marks something of a turning point in these tragic events. Thus the ground was already well prepared when Urban Hiärne, who originally had been unshakably convinced of the truthfulness of the accounts given by the kids about having been brought to the Devil by witches, in 1676 gave vent to his misgivings in a short paper "Kort betänkande om de anfäktade barnen i Stockholm" (A Short Memo About the Obsessed Children in Stockholm). Hiärne was no pioneer in the struggle against the witch hunt but his position, once he had been convinced of the facts, was clear and definitive. His intervention surely contributed to the cessation of the madness the next year so that the priests in their sermons could thank a merciful God for having delivered the land from the scourge of the witches.

Metallurgy and Spas

The belief in the health-giving effects of the water from certain natural springs must be very old indeed and was clearly widespread already in the ancient world. With the teachings of Paracelsus and his iatrochemistry followed a renewed interest in such spas on the continent of Europe as Aachen and Karlsbad, where you could take the waters and cleanse your body humors to get rid of illnesses and improve your health. In addition to a copious imbibing of the healthy water the cure at the spa also included a low diet and suitable exercise. The medical results were considered to be excellent and there was hardly any complaint that a cure at the spa could not improve. Since it was mainly the wealthy that frequented the famous spas at Aachen and Karlsbad there was money to be made in the spa business. A pretty penny was lost for the Swedish economy when the upper classes visited foreign spas and in the opinion of the enterprising Hiärne something ought to be done about this.

His knowledge of chemistry came in handy when he began to look for natural springs with water suitable for a spa in his native country. Spa water should be rich in minerals, which in reality meant sundry iron salts, and an acid component that proved to be identical to the carbonic acid discovered by van Helmont. Hiärne analysed a number of water samples from different springs without success, but finally he was lucky when water from a spring on Baron Soop's estate at Medevi in the province of Östergötland proved to fulfil Hiärne's requirements. Together with Soop he launched the first Swedish spa, which turned out to be a great success. Much to the annoyance of Hiärne, new spas appeared all over the countryside and in order to put a stop to these impudent charlatans he published a booklet where he described how the analysis of the spa water ought to be performed. However, what the criteria for proper spa water actually were according to Hiärne are not always obvious to us.

Sweden's attempt to become a great power in Northern Europe and the endless wars during the Carolingian period of increasingly absolute monarchy led to a great demand for ready money. To satisfy that demand, the export of metals such as copper and iron was crucial and that made chemistry, with its obvious importance for metallurgy, the favourite science of the Swedish sovereigns. Here again Hiärne's talents as a practical chemist served him well. He had become physician-in-ordinary to King Charles XI and chairman of the Royal Swedish Medical Board, but most important, he was head of the Royal Laboratorium Chymicum, which was housed in a newly renovated palace in Stockholm. The Royal Mining Board supervised the mining industry and the new laboratory came under this important governmental authority. In addition to the analytical work required by the Board, the Laboratorium Chymicum also produced a number of chemical compounds for pharmaceutical use, intended both for the military pharmacies and for free distribution among the poor, a doubtful benefit for soldiers as well as paupers. An early collaborator of Hiärne's at the laboratory was the outstanding German chemist Johann Georg Gmelin, who later settled in Tübingen and became the progenitor of a famous dynasty of chemists.

Hiärne was a talented analytical chemist but in his views of chemical theory he was old-fashioned and completely enthralled by the iatrochemical ideas of Paracelsus. In no way can he be compared to such a pioneering scientist as Robert Boyle, nor is his name linked to any original discovery. On the other hand, he was a good organizer and we undoubtedly owe the creation of the first advanced chemical laboratory in Sweden to him. The tradition of metallurgy and mineralogy, which Hiärne had initiated at the Laboratorium Chymicum, continued to dominate Swedish chemistry during the 18th century when it enjoyed European fame. Even after his death a number of distinguished mining engineers and chemists gave Sweden a leading position in mineralogy and metallurgy. Thus, the mining industry that had been the favourite of the Carolingian monarchs was well provided for also during the Age of Liberty that followed the death of Charles XII.

Hiärne's successor at the Laboratorium Chymicum, Georg Brandt (1694–1768), had studied medicine in Leyden under the great Boerhaave, who to everything else was also the leading chemist at the time and renewed the teaching of this subject. After having obtained his doctorate in medicine in Reims, Brandt returned to Sweden and was immediately offered the position as head of the Laboratorium Chymicum, where he was active until his death. His most important research contribution concerned the metals zinc and cobalt, the latter he actually discovered. Brandt was a good teacher and had several talented pupils. The most outstanding of them was Axel Fredrik Cronstedt (1722–1765), who introduced a new system of classification in mineralogy based on a chemical analysis of the minerals. His system became internationally accepted and was used by the leading German mineralogist Abraham Gottlob Werner. For his chemical characterization of minerals Cronstedt improved the blowpipe analysis, an old method widely used by Swedish chemists. With the blowpipe one could investigate the behaviour of a mineral when melted in a flame of different temperatures under oxidizing and reducing conditions. This technique was used extensively by later generations of Swedish chemists, including Jacob Berzelius.

The First Professor of Chemistry

Hiärne had pioneered chemistry in Sweden, but he was mainly interested in practical problems and did not involve himself in actual teaching. Instead it was Johan Gottschalk Wallerius (1709–1785) who introduced chemistry as an academic discipline at the University of Uppsala. He grew up in a vicarage in Central Sweden and after studying science in Uppsala he changed to medicine. Having obtained a doctorate he was appointed lecturer at the medical faculty. Almost immediately he became involved in a bitter conflict with no less than the young Linnaeus when they applied for the same chair in medicine at the university. Wallerius, who was a difficult and quarrelsome person, attacked his rival furiously in his lectures and was himself assailed by the adherents of Linnaeus, who kept himself aloof of the squabble in the assurance of being the favoured candidate for the chair. Of course Linnaeus was appointed professor rather than Wallerius and the disappointed candidate had to be content with his position as lecturer, in which capacity he taught anatomy and physiology. However, his scientific interests were chemistry and mineralogy and here he was so successful that when a chair was created in chemistry at the university, he was appointed to this position without any further academic conflicts.

Wallerius was hardly a very attractive figure personally and completely lacked the charm and charisma that was so characteristic of his adversary, Linnaeus, but there is no denying that he was very effective as a professor. Just four years after his appointment, the first chemical laboratory at Uppsala University was ready, built according to his own design and specifications. His main fields of research were metallurgy and mineralogy, but he was also a very competent teacher of the entire chemical subject and found time to write the first textbook of chemistry in Swedish (even if it had the Latin title *Chemia physica*). Agricultural chemistry was another subject of interest to him and he had a number of rather remarkable ideas about the nature of the soil and its importance for the growing of plants, or rather its lack of importance. The fact was that, like van Helmont and Boyle, he believed that the plants derived all their nourishment from the elements of air and water. Another of his favourite theories was that water could be transformed into soil and on prolonged distillation of water he, sure enough, obtained a residue that he identified as soil. Later it was found that the residue he took to be soil had originated from the glass-walls of the vessel used in the distillation. In conclusion, it is fair to say that even if he had nothing of the brilliance and scientific genius of his old rival, Linnaeus, Wallerius is important in the history of Swedish science as the first academic representative of chemistry.

Two Outstanding Chemists in the Era of Neoclassicism

Gustavus III is hardly known for his interest in the sciences. Even in this respect his reign implies an obvious break with the traditions from the Age of Liberty. Nevertheless, it was during this era of neoclassicism (in Sweden known as the Gustavian period, after the King) that two of the country's most outstanding chemists appeared: Torbern Bergman and Carl Wilhelm Scheele.

***

Swedish chemistry in the 18th century, notwithstanding its international reputation, particularly in mineralogy and metallurgy, had a somewhat down-to-earth character with a minimal scope for originality and personal brilliance, but Torbern Bergman (1735–1784) (Fig. 20) was the obvious exception to the rule. Here at last was a young universal genius, whose talents did not recognize any rigid borderlines between different subjects. He was the son of a royal bailiff in southern Sweden and at the age of 17 he matriculated at Uppsala University where his broad talents were quickly realized. His studies comprised about everything, from philosophy and mathematics to physics and entomology. In 1758 he submitted a thesis for a doctorate in astronomy and he then was appointed a lecturer in both experimental physics and mathematics. Bergman's main interest in the 1760s seems to have been the study of various electrical phenomena, but then something entirely unforeseen happened. In 1767 the professor of chemistry, Johan Gottschalk Wallerius, suddenly retired because of ill health (nevertheless he lived on for another 18 years). Among the four applicants for the vacant chair was Bergman. This probably makes him unique in the history of Swedish chemistry in view of his absolute lack of experience in chemistry, with the exception of a hastily compiled paper without any scientific value. His reputation as a young genius was, however, so well established that it had penetrated even to the scientifically indifferent crown prince and chancellor of the Swedish universities, the future King Gustavus III. In an inspired moment he saw to it that Bergman, to everyone's surprise, was appointed to the chair in chemistry, a discipline of which he was at that time completely ignorant.

As one might expect it did not take Bergman long to master the elements of chemistry and soon enough he advanced to become a master of analytical

Fig. 20. Torbern Bergman (1735–1784).
Courtesy of the Royal Swedish Academy of Sciences.

chemistry. He early on was interested in spas and their water and methods for its analysis, as well as the construction of apparatus for the industrial production of mineral water. The background to this interest in spas and mineral water was tragic. Bergman suffered from tuberculosis of the lungs and consumed large quantities of mineral water in a forlorn attempt to mitigate his worsening health. At the same time it gives us a harrowing insight into the therapeutic resources in those days. His analytical work with mineral water led him to attempt similar methods for the analysis of solid minerals. To make this possible he had to obtain his mineral samples in solution and he did this by treating them with suitable acids. They could then be analysed in the same way as the spa water. Based on these analyses and following the same principles as those originally laid down by Cronstedt, Bergman classified the minerals according to their chemical composition rather than their morphology (external form and appearance). This brought him in conflict with the leading mineralogist on the continent of Europe, Abraham Werner, who had once been a great admirer of Cronstedt, but now seemed to have difficulties accepting Bergman's classifications.

Bergman was no doubt hampered as a theoretical chemist by his being an unshakable adherent of the phlogiston theory. He even worked out a method for determining the amount of phlogiston in iron. In reality he had determined something entirely different, namely its carbon content. The phlogiston theory is a good example of how the building of airy theories went on even in the chemistry of the 18th century. At the same time an attempt to systematize everything in nature was also apparent. The chemists seem to have been looking high and low for principles that could be used to divide up their science into groups and classes in the same way as Linnaeus had done so successfully for botany with his classes and orders, species and genera. Bergman was of course active here and like his colleagues he discerned metals and half-metals, a great number of different "earths" and the troublesome group of "salts" that strangely enough tended to include acids and alkalis (bases). The confusing nomenclature of chemical compounds was another problem that he tried to solve. The names used for practical purposes (trivial names) often enough went back to medieval alchemy. Another legacy of alchemy was the enigmatic symbols used, for instance, to designate metals. This language of chemical symbols had become increasingly bewildering and Bergman attempted to replace it with a system of symbols that made it possible to render chemical reactions in the form of relatively simple equations. Both his nomenclature and his chemical symbols undoubtedly represented an important step in the right direction but were unfortunately never generally accepted. It would not be until the beginning of the next century that Berzelius was able to make all chemists in the world agree on a common language of chemical symbols.

In terms of chemical systematization Bergman was particularly successful with his theory of affinity. This idea about the different propensities of chemical compounds to react with each other had been advanced earlier by French chemists, but Bergman's comprehensive examination of the phenomenon, that in German came to be called "Die Wahlverwandtschaften" (approximately "Affinity Kinship"), resulted in tables of chemical affinity that aroused international interest. As an odd example of this, it may be noted that Goethe used the German term "Die Wahlverwandtschaften" as the title of a novel.

Bergman had a number of foreign pupils and collaborators, corresponded intensely with the leading chemists of the period, had his papers translated into all great European languages and was also elected a member of the French Academy of Sciences. He could certainly not complain of lack of appreciation, but his health continued to deteriorate. Even so he was perpetually active to the very last when finally his illness got the better of him and he died during a stay in the spa at Medevi in 1784.

***

The 18th century was very much concerned with the importance of scientific discoveries for the development of such economically important activities as agriculture, industry and mining. This was true also in chemistry and that period yielded a rich crop of methods for the purification of previously unknown elements and chemical compounds from natural products and their analysis. Swedish chemist Carl Wilhelm Scheele (1742–1786) (Fig. 21) is a good example of such experimentally oriented research, unencumbered by a lot of loose theoretical speculations.

It is sometimes difficult for a modern chemist to really grasp the theoretical achievements that more than anything else were the reason for Torbern Bergman's fame in his own time. In a way it feels like trying to capture a moonbeam. On the other hand, it is that much easier to understand the high esteem that Carl Wilhelm Scheele was held in both by the chemists of his own time and indeed by present-day chemists. It is the solid results of ceaseless work at the bench in his rather primitive laboratory that has formed the foundation of his fame.

Scheele was born in Stralsund, the capital of Swedish Pomerania (present-day north-eastern Germany at the Baltic Sea). He came from an originally well-to-do family, but unfortunately his father went bankrupt when the boy was only a few years old. Perhaps it was because of lack of money that Carl Wilhelm at the age of 15 was apprenticed to an apothecary in Gothenburg instead of being given the academic education that his talents undoubtedly qualified him for. Undaunted, he studied chemistry by himself in his free time and fortunately enough his benevolent

Fig. 21. Carl Wilhelm Scheele (1742–1786).
Courtesy of the Royal Swedish Academy of Sciences.

employer permitted him to make experiments in the laboratory of the pharmacy. Having advanced to journeyman he moved to Malmö in southern Sweden, where he was employed by yet another friendly pharmacist, who also let him use the localities of the pharmacy for the experiments that he devoted all his free time to. Here he made the acquaintance of Anders Retzius, a young assistant professor at the nearby University of Lund. Retzius proved to be very important for the future scientific career of the self-taught chemist. After a couple of years in Malmö and a relatively short stay in Stockholm, he found himself in the pharmacy Uplands Vapen (Arms) in Uppsala where he remained for five happy and scientifically very productive years. While in Stockholm he tried to publish a paper in the *Proceedings* of the Academy of Sciences but the editor, none other than Torbern Bergman, rejected the paper, a very unpromising beginning of what would later turn out to be a warm friendship and a very fruitful scientific collaboration. In any case, two years later a paper by Scheele appeared in the *Proceedings* even if it had to be submitted by his friend Retzius. It would be the first in a long series of scientific publications.

The five years he spent at the pharmacy in Uppsala were probably his most successful scientifically and his working capacity was truly amazing. However, he needed a more permanent position and a better income than Uppsala could offer him. When it became known that the local pharmacist in the little town of Köping had recently died and that the young widow was prepared to transfer the privilege as apothecary of her late husband to Scheele, who was by now well known as an outstanding young chemist, he made up his mind to move to Köping. Here he remained until the end of his life, incessantly occupied with experiments in his laboratory, originally located in a shed in the backyard, but later moved into the localities of the pharmacy itself. He seems to have been happy as a pharmacist in Köping and his relations with the young widow were also quite amiable. Not even an attractive offer from none less than Frederick II of Prussia to become professor of chemistry in Berlin could tempt him away from his abode in Köping. However, he was not destined for a long life and in the autumn of 1785 he became ill with diffuse symptoms, which were put down to gout, at this time a diagnosis used also for all kinds of rheumatic complaints. He continued to work in his laboratory but his health deteriorated steadily and in May 1786 he died, after having on his deathbed married the faithful widow and transferred the privilege as apothecary to her.

Scheele in every way agrees with the conventional picture of a young chemical genius, working incessantly among the flasks and test tubes in his laboratory where he constantly discovers new unknown chemical compounds. His gluttony for work is borne out by the copious laboratory journals that he left behind and that were not systematically examined until the end of the 19th century. He discovered the element of chlorine by treating manganese superoxide (peroxide) with hydrochloric acid and he also managed to produce pure manganese. Here he worked closely with a pupil of Torbern Bergman's and generally speaking there was an intimate collaboration between the two leading Swedish chemists once the discontent of their first encounter had evaporated. He also prepared hydrofluoric acid from calcium fluoride and he was the first to purify arsenic acid, molybdic acid and tungstic acid. He was furthermore active in organic chemistry, at this time a field that was fairly unexplored, and here he for instance discovered oxalic acid, lactic acid and uric acid. He even ventured into biochemistry and tried to characterize extremely complicated substances such as proteins, which he showed to contain sulfur.

Of course it is his discovery of oxygen that forms the foundation of his international fame and there exists a whole literature about the priority here. Does it belong to Scheele or to English amateur chemist Joseph Priestley (1733–1804)? There can be no doubt that Priestley produced oxygen in 1774 by heating mercuric oxide and that he published his findings the next year. Scheele, on the other hand, did not publish his discovery of oxygen until 1777 in his book *Chemische Abhandlung von der Luft und dem Feuer* (A Chemical Treatise of Air and Fire). However, when his laboratory journals were thoroughly examined in 1892, it became apparent that he had produced oxygen already in 1772 and demonstrated its ability to sustain combustion. In a letter to Lavoisier in 1774 he described his work with oxygen (he called it "vitriolic air") and it would seem reasonable to conclude that Priestley and Scheele discovered oxygen independent of each other. In any case, Scheele's contributions to chemistry are solid and long-lived, not being encumbered with loose theoretical speculations, as was so often the case in the 18th century.

Berzelius Takes Charge

With the untimely deaths of both Torbern Bergman and Carl Wilhelm Scheele, the Academy of Sciences had lost much of its research potential in the sciences. A period of relative inactivity followed. It was after all a royal academy but the King Gustavus III had little interest in the sciences and his successor Gustavus IV Adolphus was probably the most incompetent of Swedish kings. The support that the Academy had previously enjoyed in the highest circles of Swedish society, ranging from King and nobility to the great financial magnates, seemed to have evaporated. What was needed in order to raise the Academy to its former grandeur was clearly a scientist with something of the same international fame as Carolus Linnaeus. Fortunately, such a man made his appearance in the first decade of the 19th century: Jacob Berzelius (1779–1848) (Fig. 22).

Fig. 22.   Jöns Jacob Berzelius (1779–1848).
Courtesy of the Royal Swedish Academy of Sciences.

He came from a family of clergymen. After the early death of his father Jacob's mother remarried the Rev. Anders Ekmarck, a widowed vicar in the countryside of southern Sweden with five children of his own. Here the boy spent a reasonably happy childhood until the death of his mother forced the stepfather, who seems to have been fond of the boy, to reluctantly place him in the care of a maternal uncle with an alcoholic wife. The years that Jacob spent with this new family were obviously miserable and it was a relief to him when he was old enough to be admitted to the gymnasium in the nearby town of Linköping. However, his schooldays would not turn out to be without problems.

The swotting of Latin and Greek in the schools of those days was not to the liking of Jacob Berzelius and particularly when the weather was fine he much preferred to wander around in the countryside with a shotgun. On such a day he happened to run into the headmaster who immediately forbade such extracurricular activities. When he caught Jacob out again hunting a couple of days later the headmaster was enraged and sentenced the disobedient schoolboy to a birching and, even worse, he threatened to expel him from the school. Fortunately for Jacob, one of the senior teachers took a more lenient view of his offence and intervened on Jacob's behalf with the bishop, who as the inspector of the school let the culprit off with an admonition. However, the insulted headmaster had his revenge when Berzelius' leaving certificate from the gymnasium described him as a young man of good natural talents, but with unsatisfactory morals and a questionable future.

Nevertheless, Berzelius was admitted to medical school at Uppsala University where he discovered that chemistry held a special fascination for him. Having submitted his rather undistinguished thesis he got an unsalaried position at what would a few years later become the Karolinska Institutet in Stockholm, where he was entrusted with teaching future military surgeons the elements of what we today call biochemistry. He soon discovered the urgent need for a textbook in this newfangled subject and in 1806 he published the first volume of his classical book *Föreläsningar i Djurkemien* (Lectures on Animal Chemistry). This brought him to the attention of the authorities and the next year he was appointed professor of chemistry and pharmacy. His duties as a teacher at the medical school may have been mostly concerned with animal chemistry but his research was increasingly directed at inorganic chemistry, which was much more amenable to analysis that could give definitive and reproducible results.

We have already briefly considered some of Berzelius' major scientific achievements but we must now discuss his role as one of the leading European chemists and the permanent secretary of the Royal Swedish Academy of Sciences. He had been elected a member of the Academy in 1808 when he was 29 years old.

The young chemist had undoubtedly set his goal high not only in his research but also in his ambitions as a rising star in the Academy. Even as a junior member he was determined to become its permanent secretary and eventually raise it to its former glory in the days of Linnaeus. However, as it turned out he had to wait ten years before he reached this elevated position.

International Contacts

Berzelius' thesis, which dealt with the effect of electricity from the pile of Volta on patients at the Medevi spa, has certainly left no traces in the history of medicine. However, a year later, having made the acquaintance of the wealthy mill-owner Wilhelm Hisinger at the Galvanic Society in Stockholm, he and the scientifically very interested Hisinger began to investigate the behaviour of salts in water solution under the influence of an electric field. They discovered that what we today call negative and positive ions moved to different poles in the electric field, the negative ions to the positive pole and the positive ions to the negative pole. When these results were published in 1803 they awakened international interest and eventually brought Berzelius into contact with English scientist Humphry Davy (1778–1829) (Fig. 23) who had investigated similar problems.

Davy had at an early age made a name for himself through his youthful energy and enthusiasm in combination with an uncanny scientific intuition, while Berzelius was still fairly unknown. Nevertheless, their studies of electrolytic processes were clearly conducted independent of each other. In 1807 Davy made an important discovery when he examined the behaviour of slightly moistened potash (potassium carbonate) in an electric field. Davy found that a silvery matter was deposited on the negative pole, while oxygen was liberated at the positive pole. He concluded that the silvery deposit at the negative pole was a metal, which he named potassium. In similar experiments including sodium hydroxide and several of the so-called alkaline earths, he was able to isolate sodium, magnesium, calcium, strontium and barium. Another important discovery of Davy's concerned what Lavoisier had called muriatic acid (hydrochloric acid) and believed to contain oxygen. Davy could show, however, that muriatic acid did not contain oxygen and was in reality what we now call hydrochloric acid. He also characterized the element chlorine (previously believed to be a higher oxide of muriatic acid).

Davy enjoyed something of a meteoric career. He was elected a fellow of the Royal Society in 1803 and in 1812 he was knighted. In 1818 he was made a baronet and in 1820, on the death of Sir Joseph Banks, he succeeded him as president of the Royal Society. When he and Berzelius finally met during a visit of the latter to England in 1812, Berzelius was rather irritated by what he considered to be the

Fig. 23.   Humphrey Davy (1778–1829).
Picture from the Internet.

haughty attitude of Davy and they never became very friendly with each other. Otherwise, it would seem that Berzelius had the gift of friendship with an extensive net of scientific relations both in England and on the continent of Europe. Over the years many foreign scientists, particularly from Germany, worked in Berzelius' laboratory and considered themselves his pupils.

In the long line of collaborators from abroad that came to work with Berzelius, two were particularly close to the heart of the old master, Eilhard Mitscherlich (1794–1863) and Friedrich Wöhler (1800–1882). Mitscherlich had originally been a humanist specializing in oriental languages and Persian history, but had changed to chemistry and made an important discovery, the phenomenon of isomorphism. This means that chemical compounds, although they have different structures, may, nevertheless, show the same crystal form. Berzelius had been very impressed by the young linguist-turned-chemist and recommended him for the chair in chemistry in Berlin that had recently become vacant after the death of Martin Klaproth. In order to improve Mitscherlich's chances to obtain the professorship Berzelius invited him to Stockholm for a period of two years. They became close friends and Berzelius always tried to promote Mitscherlich's scientific career.

However, among all the young German chemists that Berzelius came into contact with, Friedrich Wöhler had a special position. In his memoirs *Jugend-Ehrinnerungen* (Memories of My Youth), Wöhler gives an animated account of his first meeting with Berzelius in 1823 in Stockholm. Wöhler had looked forward to this encounter with great expectations but also with some nervousness, since Berzelius was at the peak of his international fame at the time. As it turned out, their personalities seem to have been perfectly matched and this was the beginning of a lifelong friendship. Wöhler put in an enormous amount of work translating Berzelius' publications into German and undoubtedly his efforts contributed considerably to the dominance of Berzelius in the German-speaking countries.

Berzelius probably exerted his greatest influence through his Annual Reports (*Jahresberichte*), which he published in German from 1822. Here he freely supported his scientific favourites, for instance Mitscherlich, and at the same time he did not pull his punches when criticizing his opponents. There can be no doubt that these Annual Reports played a considerable role in restoring to the Swedish Academy of Sciences something of the standing that it had enjoyed in the days of Linnaeus, Bergman and Scheele. Another important circumstance was the fact that its permanent secretary, Berzelius, was able to place his own admiring pupils in leading academic positions, particularly in German-speaking countries. Nor did it harm the Academy that the King, Charles John XIV, formerly one of Napoleon's marshals, who had made a complete about-turn and now strongly supported the enemies of his previous master, was a great admirer of Berzelius. In fact, the royal princes attended the courses given by Berzelius and regularly worked in his laboratory. Chemistry had become something of a fashionable science so that former cabinet ministers of the highest aristocracy, such as Count Hans Gabriel Trolle-Wachtmeister, were among its students and intimate friends of Berzelius'. Even Linnaeus himself had not been able to attract so much royal interest.

Berzelius undoubtedly had strong views on his own field, chemistry, but also on science in general and relevant aspects of philosophy. In his youth, particularly in his well-known *Lectures on Animal Chemistry* of 1806, he proclaimed himself in no uncertain terms as a follower of what was then called "atomism" or "materialism". According to Berzelius, "life or vital force is not something indefinite, a foreign essence deposited in a living body that makes inanimate matter come alive. Organic nature therefore obeys the same chemical laws as inorganic nature. Vital force is merely a name for the sum of the chemical and mechanical processes in the living body." However, Berzelius was not consistent in his rejection of vital force as an explanation of life as different from inanimate matter. Later on, when he stood on the peak of fame and prestige, he gave vent to opinions that came close to vitalism. On the other hand, at the end of his life he would seem to have returned to the position of his youth: "Vital force is an unnecessary and detrimental assumption; organic processes obey the same laws as inorganic ones."

If Berzelius seems to have vacillated frequently between atomism/materialism and vitalism, he always rejected natural philosophy as implying "ignorance of everything factual, love of poetry and the fine arts, and a trusting and unthinking devotion to personal opinions, which through their incomprehensibility have gained a reputation of being profound." In a letter to his friend, the famous botanist Carl Adolph Agardh, who had fallen victim to natural philosophy and allowed that to influence a textbook he had just written, Berzelius admonished him: "The truth, my friend, is the soul of science. She cannot tolerate even the dreams of the genius."

In his opinion of natural philosophy and the effect of romanticism on scientific thinking, Berzelius was eminently discerning and reasonable, but he could also sometimes cling stubbornly to his own pet theories even when they were contradicted by experimental facts. For instance, he had great difficulties accepting that muriatic acid was really hydrochloric acid and did not, as Lavoisier had claimed, contain oxygen. However, the best example of Berzelius' stubbornness is his conflict with French organic chemist Auguste Laurent about the possibility of replacing a hydrogen atom bound to carbon with a chlorine atom (substitution). According to Berzelius' view of the chemical bond, it was always dependent on an electrostatic attraction between particles (atoms) with opposite electrical charges.

Consequently, hydrogen being electropositive could not be substituted by chlorine, which was electronegative. The substitution reaction that Laurent had reported must simply be a mistake! Of course, Berzelius was wrong here and the way he stuck to his guns and refused to accept the facts was in the end detrimental to his reputation as the leading European chemist of the time.

Of Minerals and Catalysis

Linnaeus had based his sexual system for the classification of plants on their exterior characteristics; it was a purely morphological system. Similar principles can be said to have dominated the attempts to obtain a system for the classification of minerals in the 18th century. German geologist Abraham Gottlob Werner (1749–1817) used mainly such properties of minerals that could be observed by visual inspection for his taxonomic system. Chemical analysis was considered to be uncertain and therefore of limited usefulness in mineralogy. However, a Swedish school objected to such fixation on the appearance of the minerals and instead stressed the importance of their chemical composition as the basic taxonomic principle. The leading names of this chemical school were Torbern Bergman and the outstanding mining expert Axel Fredrik Cronstedt (1722–1765), who put little stock in the appearance of a mineral as the basis of classification. A weakness of the chemical school was that they did not realize the principle later established by Proust that the elements present in a chemical compound, which has been obtained in pure form, always appear in the same proportions (the law of definite proportions). French abbé and mineralogist René Just Haüy (1743–1822) can be said to represent something of a middle course since his school of crystallography took both the appearance and the chemical composition of a mineral into account as the basis of its classification. Unlike both Bergman and Cronstedt he was aware of the importance of definite chemical proportions.

It would take some time before Berzelius became interested in minerals as something more than just a source of new elements to be discovered. Perhaps his friendship with Johan Gottlieb Gahn (1745–1818), the venerable old chemist, mill owner and mining expert, had something to do with the fact that around 1815 he decided to take up mineralogy. Of course he adhered to the principles of Bergman and Cronstedt where chemical analysis was the basis of taxonomy. Being a master of purification and analysis gave him a definite leading edge here. It is enough to mention his elucidation of the role of silicon compounds in a number of minerals to illustrate this. As might be expected, the old-fashioned morphologists of the Werner school indignantly dismissed Berzelius' system as worthless since it did not take the exterior appearance of the mineral into account. However, later on

Berzelius came to agree with Haüy's compromise that both the morphology and the chemical composition were of importance for the classification. In any case, there can be no doubt that he played a leading role in the introduction of modern chemical principles in mineralogy. In addition to this, Berzelius' intimate relations with such leading names in the Swedish mining industry as Hisinger and Gahn had early on given him easy access to all sorts of minerals. Already in 1803 he and Hisinger discovered the element cerium, which resulted in a conflict with the famous German chemist Martin Heinrich Klaproth, who had made a similar discovery. In 1817 Berzelius discovered the element selenium and in 1827 he found thorium.

It has often been said of Berzelius that his greatness as a scientist lay not in his ability to come up with new ideas and stimulating hypotheses, but there is an important exception to this assertion. In his Annual Report of 1835 he introduced the concept of catalysis, one of the most fundamental of chemical principles. He gave an account of a number of catalytic phenomena from inorganic chemistry known at this time, for instance the effect of metallic platinum on the combustion of hydrogen in the presence of oxygen. Furthermore, he pointed to the discovery of Kirchhoff that acids can convert starch into glucose without themselves being consumed in the reaction. He then went on to give the now classical definition of the terms "catalysis" and "catalyst". Berzelius emphasized that a catalytically active substance can rouse "affinities" that are normally "asleep" at a certain temperature, by its mere presence. He compared the effect of a catalyst to that of heat, which can speed up chemical reactions, but he suggested that a catalyst unlike heat is specific and only affects certain reactions. He then discussed catalytic processes in living organisms and took the presence of diastase around the "eyes" of a potato as an example. Finally he drew an almost visionary picture of how thousands of catalytic processes in the living organism help to convert nutrients into the innumerable compounds which make up the cell.

In a way none of this is new; already the old alchemists thought in terms of "ferment" and "fermentation". However, such terms meant something entirely different in those days. When Berzelius introduced his concepts of catalysis and catalysts, even the great German chemist Justus von Liebig (1803–1873) believed that a sugar molecule which was being fermented to alcohol and carbon dioxide could transfer the process of fermentation to other sugar molecules in the vicinity, "infect them with fermentation" as it were, by way of some mysterious molecular vibrations. It was not until Berzelius defined catalysis and catalysts that concepts such as ferments and fermentation acquired intellectually satisfying, stringent meanings.

An Unexpected Responsibility

In the 18th century the creation of the Swedish Academy of Sciences with names such as Linnaeus, Bergman and Scheele had played a decisive role in the development of science in Sweden and under the leadership of Berzelius the Academy continued to be very influential. Berzelius was a great traveller even in his old age when he was not in the best of health. There is no doubt that these personal contacts with leading chemists on the continent of Europe, but also in Great Britain, were important not only for his own research but also for the Academy and its international standing. When he died in 1848 it meant the end of an epoch and the Academy seemed destined to remain a national scientific academy among many others in Europe, without any distinguishing features. That changed, however, when Alfred Nobel in his will of 1895 made the Royal Swedish Academy of Sciences responsible for two of the Nobel Prizes, those in physics and in chemistry. This led to a complete change of focus in the activities of the Academy. Strangely enough, in the years between the death of Berzelius and the advent of the Nobel Prizes, the Academy had seemed to be heading in an entirely different direction, towards polar expeditions and arctic research.

Ventures Into the Arctic

What followed was a period when the efforts of the Academy were to a large extent concentrated on the spectacular mission of polar exploration, an activity that at least in the eyes of the public was considerably more fascinating and easy to understand than for instance physics and chemistry. Thus the Academy was involved in all the major Swedish expeditions to the Arctic area, even if much of the financial support came from private donations. However, from a scientific point of view the Academy had a leading role and could be seen as the guarantor of the quality of these projects.

Previously Russia in the first half of the 18th century had mounted expeditions led by Danish explorer Vitus Bering, hoping to find a sea route from the Arctic to the Pacific Ocean, thereby establishing a new way to America, the Northeast Passage. The first half of the 19th century had seen a number of attempts by the British to find what was called the Northwest Passage, i.e. a sea route north of Canada leading from the Atlantic to the Pacific Ocean. On the whole these attempts had been failures and in some cases, for instance the expedition led by Sir John Franklin, had resulted in a complete catastrophe with the loss of the whole expedition.

European nationalism had its heyday in the 19th century and the outburst of polar expeditions can in fact be seen as a peaceful way of asserting ambitions that would later on lead to the World War catastrophes of the next century. However, the reason for the growing Swedish interest in exploring the Artic regions was peaceful enough and originally related to the old debate about the age of the earth. Earlier attempts to calculate its age based on the teachings of the Bible had given results such as those proposed by Bishop Ussher. According to him the world had been created as recently as 4004 B.C. On the other hand, the results of Georges Cuvier and other leading palaeontologists clearly indicated that the earth was very much older than that.

The study of its fossil fauna raised the question of how these innumerable fossil organisms had become extinct. Lamarck simply denied that any extinction had occurred; it was instead a question of organisms changing in appearance over time beyond recognition. Against Lamarck's evolution Louis Agassiz maintained his theory of extinction through a catastrophe with the appearance of a new and more progressive fauna to take the place of the extinct one. Charles Lyell, on the other hand, believed in a more gradual process where extinct species were replaced by new organisms.

Of special interest to the countries of Northern Europe was the glacial theory proposed by Louis Agassiz. The Nordic hemisphere had been covered by an enormous inland ice, which had led to the extinction of all life. The present-day glaciers in the Alps and in Northern Europe, including Greenland, were remains of this presumed continental ice sheet. A number of objections had been raised against Agassiz's glacial theory as being too fantastic, but there was also growing evidence in support of it, not least through the efforts of Swedish polar explorers. Sven Lovén had been professor at the Museum of Natural History in Stockholm and was a member of the Academy from 1840 until his death in 1895. This pioneer of polar research in Sweden had in 1837 travelled to Spitsbergen in order to study the Arctic fauna. Having returned home he examined fossil snails and mussels in Sweden and found that they belonged to species that nowadays live only in the northernmost parts of the Arctic. He came to the conclusion that Sweden must at some time have been covered by a land ice that he thought could have been several hundred metres thick.

One of Lovén's pupils, Otto Torell, had been intrigued by a fossil form of a particular mussel in Sweden closely related to organisms that now lived only in the Arctic area. He had inherited a small fortune from his father and he decided to spend it on a series of polar expeditions in order to find concrete evidence for the glacial theory of Agassiz. In 1858 he travelled to Spitsbergen accompanied by a young scientist, Adolf Nordenskiöld. They studied the flora and fauna of the island and the next year Torell defended his thesis on the molluscs of Spitsbergen.

For his next expedition he had the support of the Academy of Sciences that helped with the raising of funds, including financial support from both the Riksdag and the royal house. The expedition that counted 26 members sailed to Spitsbergen in two vessels specially designed and fitted out for travelling in the Arctic. It resulted in a considerable collection of both zoological and botanical specimens as well as a number of geological observations. Their findings were consistent with the assumption that the whole of Scandinavia had been covered by a thick sheet of ice, which had played a major role in the shaping of the landscape, including the deposition of the mysterious "erratic" boulders over its surface. There is no doubt that Torell provided real proof of the glacial theory that Agassiz had outlined in 1837.

Nevertheless, the leading role in Swedish Arctic research was rapidly taken over by Adolf Nordenskiöld, who was a very ambitious and highly competent explorer and at the same time a brilliant organizer, while Torell had been above all the great enthusiast and inspirer. A new polar expedition was mounted in 1864, again supported by the Riksdag and the Academy but this time Nordenskiöld was put in charge, not Torell. In fact, Torell was not even allowed to participate unless he himself was prepared to foot the cost of his journey. It would seem that for some reason certain leading members of the Academy had formed a low opinion of him as the leader of polar expeditions. There is no doubt that Torell felt the slight very keenly and in spite of many efforts on his part he was never given the chance of another Arctic exploration. However, even if the Academy's treatment of Torell seems mean and shabby, it is hard to fault their choice of Nordenskiöld as leader of the expedition in view of the brilliant success of both this and many following journeys.

The expedition of 1864 to Spitsbergen gave very satisfactory results in terms of cartographic work and the collecting of geological, botanical and zoological material. On the other hand, it had not been possible to advance further north from Spitsbergen towards the pole. Private donors such as merchant Oscar Dickson in Gothenburg financed a later expedition in 1868. The expedition ship reached the record latitude of 81° 41′ N, but at the same time Nordenskiöld became convinced that the solid ice which covered the Arctic Ocean made it impossible to reach the pole in a ship. In the early 1870s he made an attempt, again financed by Dickson, to reach the pole by way of Greenland, travelling on foot over the inland ice. The expedition to Greenland was very successful in terms of collecting scientific material but at the same time those years spent on the ice highlighted the enormous difficulties of further progress towards the pole. Consequently Nordenskiöld decided to concentrate instead on another classical goal in Arctic exploration: finding the Northeast Passage.

Economic support of this new project came as before from the Riksdag and King Oscar II, but above all from private donors with Dickson as the most important contributor. The seal-hunting vessel Vega was specially fitted out for an Arctic expedition and in the summer of 1878 it left Tromsø on the northern coast of Norway and set out for the mouth of the river Yenisei and from there to the Bering Strait where the Vega as planned took up winter quarters. When the ice broke in July the ship could continue its journey through the Bering Strait, thereby completing the first voyage through the Northeast Passage. The journey home was an unprecedented triumphal progress and on the ship's arrival in Stockholm the enthusiasm of the public knew no bounds. Vega's venture through the Arctic Ocean to find the Northeast Passage to the Pacific aroused the whole Swedish nation to a patriotic fervour and some of the glory probably spilled over to the Academy of Sciences where much of the planning and organizing had after all taken place.

The Advent of Ions

When Berzelius vacated his chair in pharmacy and chemistry at the Karolinska Institutet he was succeeded by one of his pupils, Carl Gustaf Mosander (1797–1858), whose research dealt mainly with the discovery and characterization of new elements. Thus he can be said to have continued an important line of research inherited from Berzelius. In fact, Mosander demonstrated the existence of more elements than his old master, even if he could only obtain a few of them in pure form. When we try to judge his contributions to inorganic chemistry it is important to remember that a major analytical method such as spectroscopy did not become available until the 1860s. Against this background his work on the rare earth group of elements and the isolation of lanthanum and cerium as pure metals is indeed impressive, in spite of the fact that cerium had been discovered as an oxide by Berzelius and Hisinger in 1803. It is therefore fair to say that even after the death of Berzelius the chemical research done at the Karolinska Institutet was still of a high quality and attracted international attention. The same is true of Uppsala with names such as Anders Ångström, Olof Hammarsten and Per Cleve. Nevertheless, one has the same feeling as after the demise of Torbern Bergman and Carl Wilhelm Scheele of Swedish chemistry waiting in rapt expectation for someone to appear who could restore it to its former glory.

The history of science has many examples of late bloomers who took their time before they were able to demonstrate the extraordinary talents that would ultimately lead to remarkable research careers. Names such as Jacob Berzelius, Louis Pasteur and Claude Bernard come to mind here. On the other hand, there were also young geniuses such as Albrecht von Haller, Torbern Bergman and Antoine-Laurent Lavoisier who early on were recognized as being extremely gifted. In this latter category we must include Svante Arrhenius (1859–1927) (Fig. 24) who entered

Fig. 24.   Svante Arrhenius (1859–1927).
Courtesy of the Royal Swedish Academy of Sciences.

Uppsala University at the age of 17. He chose physics as his major subject but the wilful young student soon came into conflict with the professor of physics, Tobias Thalén, who flatly refused to accept any student who wanted to work on a thesis outside the professor's favourite research subject, spectroscopy. When Arrhenius would not budge from the subject he himself had chosen for his thesis, "The Galvanic Conductivity of Electrolytes", there was nothing for it but to transfer his studies to the department of Professor Erik Edlund, who was a physicist at the Swedish Academy of Sciences in Stockholm. Here Arrhenius found a much more congenial atmosphere and in 1883 he completed the experimental part of his thesis. On the advice of the chemistry professor at the Stockholm Högskola, Otto Pettersson, who would become a close friend and later colleague, he extended the theoretical part of the thesis under the subtitle "Theoretical Chemistry of Electrolytes", something that made it even less acceptable to Professor Thalén, who was after all supposed to be Arrhenius' thesis adviser.

Nevertheless, he submitted his thesis in Uppsala in May 1884, where it was grudgingly accepted but given the fairly low mark of *non sine laude approbatur* (approved not without praise) while his defence, which had been brilliant, received a mark of *cum laude approbatur* (approved with praise). This low mark meant that he was not considered qualified for a docentship, which was a prerequisite if he wanted to apply for an assistant professorship. It was a blow not only to his future career but also to his self-esteem and Arrhenius never forgave "the big noises in Uppsala", as he used to call them. As we shall see in the following, other scientists both in Sweden and abroad had an entirely different and much higher opinion of his work.

What was the gist of the thesis that the misguided professors on the examining board had so completely failed to appreciate? In the experimental part of his thesis Arrhenius examined the effect on the molecular galvanic conductivity when already highly dilute solutions of electrolytes in water were further diluted. He found somewhat unexpectedly that such extreme dilution in the majority (but far from all) of the electrolytes investigated led to small but reproducible increases in the molecular conductivity. For his negative results, where dilution gave a lower molecular conductivity, it was easy enough to find trivial explanations that had to do with the quality of his distilled water, but what about the cases where the conductivity increased? It seemed to Arrhenius that the present theory about the conductivity of electrolytes had to be revised in order to accommodate his new data.

He made the assumption that a water solution of an electrolyte contained both "active" molecules capable of conducting electricity and "inactive" molecules that did not have this ability. Furthermore, active and inactive forms of the electrolyte were in chemical equilibrium with each other so that at infinite dilution all electrolyte molecules were in the active form. He also assumed that it was the active molecules which participated in the chemical reactions the electrolytes could undergo. It may seem obvious to us with the benefit of hindsight that Arrhenius must have realized the true explanation of his results already when he wrote his thesis. It was because of their dissociation into free ions that the active molecules showed electric conductivity as well as chemical reactivity! In fact, on several occasions later on Arrhenius would claim that he had this insight already in 1884. However, on balance it appears more likely that it was during his so-called *Wanderjahre* (1886–1890) that he came to this conclusion. In order to understand the importance of this period in the life of the young Arrhenius we must consider how his theories were received in the world outside Uppsala. We only have to move from the rigid conservatism of the old university to the newly established Stockholm Högskola, with its more modern ideas of the fruitful borderland between physics and chemistry, to encounter a very different view of Arrhenius' thesis.

Professor Otto Pettersson had been in close contact with Arrhenius when the theoretical part of the thesis was being prepared and now he wrote a very appreciative review of the dissertation, in which he took exception to the opinion of the faculty: "The faculty have given the mark *non sine laude* to this thesis. It is a very cautious but also very unfortunate choice of mark that they have made. One can make serious blunders from pure cautiousness. There are in fact chapters in Arrhenius' thesis which are worthy of the highest praise that the faculty could bestow on it." Perhaps even more important, his work had been recognized as highly interesting by young scientists in Europe who were pioneers in what we today call physical chemistry. Arrhenius had sent his thesis to Wilhelm Ostwald (1853–1932) who was at the time professor at the Technical University of Riga and an expert on chemical affinity. He immediately realized that Arrhenius' work was of the greatest interest to his own research and in the summer of 1884 he went to Uppsala to meet the author of the thesis in person. They hit it off marvellously and remained great friends for the rest of their lives. Before Ostwald left Sweden he promised Arrhenius a position in Riga if nothing should be forthcoming in Sweden. At the same time he invited Arrhenius to go to Riga as soon as might be. The next year on the death of his father Arrhenius realized that he could now afford a couple of *Wanderjahre*, in particular since the Academy of Sciences had awarded him a travelling stipend. In the spring of 1886 he set out for Riga and then followed a European travelling period of four years, during which he visited and worked in a number of laboratories and came to know a great many scientific colleagues and, incidentally, together with Ostwald and Jacobus Henricus van't Hoff (1852–1911) laid the foundations of physical chemistry.

Of Arrhenius' *Wanderjahre* the year 1887 is particularly important in the triumphal progress of physical chemistry. In fact, it has been called its *annus mirabilis*. During this year Ostwald completed his seminal textbook *Lehrbuch der allgemeine Chemie* (Textbook of General Chemistry) and together with van't Hoff started the important *Zeitschrift für Physikalische Chemie*, which came to be the leading journal in this field, and last but not least, Arrhenius' theory of electrolytic dissociation was published in its final form, even if part of it had appeared already in his thesis. To boot, Ostwald was called to a chair in physical chemistry at the University of Leipzig (a professorship at a German university had always been his ambition). Realizing that by introducing the newfangled physical chemistry on an equal footing with the dominant organic chemistry they had turned chemistry upside down, van't Hoff wrote in a state of alarm to Ostwald: "May God have mercy on our souls!"

Jacobus Henricus van't Hoff (Fig. 25) was originally an organic chemist who in 1874, at the early age of 22, had defended a fairly undistinguished thesis at the University of Utrecht. The same year he published a short essay on the stereoisomerism of carbon compounds, assuming that the four valences of the carbon atom were pointed from the centre of a regular tetrahedron in the direction of its corners. Originally written in French it was a year later translated into German. When it came to the attention of the famous and terribly temperamental organic chemist Hermann Kolbe, he wrote a review of it that foamed with rage. This outburst made people interested and van't Hoff's essay became widely read and in fact highly appreciated by such luminaries of organic chemistry as Emil Fischer and Adolf von Baeyer. In this way van't Hoff came to be famous as the creator of a new and important concept, stereochemistry. However, his road to fame was full of stumbling blocks and it was not until 1879 that he was offered a chair in chemistry in Amsterdam.

In the 1880s he became interested in chemical equilibrium and the diffusion of solutes across semipermeable membranes, which permitted solvents like water to pass but not the solute. What was the affinity between solute and solvent and how could it be measured? Plant physiologist Wilhelm Pfeffer had come up with the idea of trying to stabilize the extremely fragile membranes formed in layers between the solutions of different inorganic substances. By letting the membranes be formed in walls made up of porous porcelain they became stable enough to make it possible to measure the osmotic pressure directly with a manometer. With this setup he had measured the osmotic pressure of a 1% sucrose solution in water and reported it to be 2/3 atmosphere.

By theoretical considerations van't Hoff realized that the osmotic pressure of a solution could be calculated from the freezing point depression caused by the dissolved solute. The osmotic pressure that he could calculate for sucrose solutions

Fig. 25. Jacobus Hendricus van't Hoff (1852–1911).

From Les Prix Nobel en 1901.

agreed well with those measured by Pfeffer. It seemed that the osmotic pressure (P) of different solutes was related to the volume (V) and the temperature (T) in the same way that was known to be the case for a mole of any gas: $PV = RT$, where R is a constant that all gases have in common. For a mole of any compound in solution van 't Hoff wrote the following instead: $PV = iRT$, where i is a measure of the deviation from the gas law. In many cases, for instance sucrose solutions, the value of i is close to one but often enough, notably for electrolytes, i is considerably higher than one. In 1885 van't Hoff submitted three manuscripts about his findings to the *Proceedings* of the Swedish Academy of Sciences and he also sent them to Svante Arrhenius to have his opinion. Arrhenius almost instantly realized that the high osmotic pressure of electrolytes, which van't Hoff had not been able to explain, could be understood if one assumed that they were dissociated into free ions that also were osmotically active. In 1887 he published his electrolytic dissociation theory in the *Proceedings* of the Academy. This was not a widely read journal but when Ostwald saw to it that both van't Hoff's and Arrhenius' theories were published in their definitive form in the first number of *Zeitschrift für Physikalische Chemie*, they rapidly became known to the scientific community worldwide, while the journal itself from that day was recognized as the leading in the field.

Alfred Nobel and His Prizes

On the face of it one would not expect that a man, who had made a vast fortune on the invention of explosives such as dynamite and whose patents also included such warlike items as smokeless gunpowder, should at the end of his life write a testament where he gave all his money to a number of prizes, which rewarded not only scientific discoveries and literature but also efforts to promote peace between nations. Nobel passed away on 10 December 1896, after which followed years of wrangling between his long-suffering but tenacious testament executors and some of his relatives, who wanted to contest the will. When that conflict was finally settled and Nobel's international assets had been successfully transferred to Sweden there still remained a number of thorny problems. The Nobel Foundation had to be created with the commission to administer the enormous fortune, the proceeds of which were to be given out as yearly prizes.

A major difficulty was that most of the institutions that Nobel had named as custodians of the Nobel Prizes in his will, with one exception, raised a number of objections when they were approached by the executors. Nobel had wanted to entrust the responsibility for the Peace Prize to the Norwegian Storting (Parliament) and unlike the other institutions named in the will it made no difficulties and gladly accepted its new duties. He had selected the Swedish Academy, an academy of letters established in 1786 by King Gustavus III, as the awarder of the Prize in Literature. After some initial procrastination by certain members the influential permanent secretary of the Academy managed to obtain a solid majority for the proposal. Regarding the Prize in Physiology or Medicine, the Karolinska Institutet declared itself ready to assume the task of awarding it, provided that certain changes were made to specifications in the will.

The real difficulty was with the Royal Swedish Academy of Sciences, which Nobel had charged with the responsibility for two prizes, in physics and in chemistry. Unless the Academy of Sciences declared itself willing to take on these duties, the whole Nobel project was in danger of collapsing. Nevertheless, the Academy dragged its feet. Not until a final agreement with the Swedish Nobel relatives had been successfully negotiated and after much haggling in plenary sessions with the whole Academy did the august body agree to undertake the task of selecting the laureates in physics and chemistry. It was now possible to set up the Nobel Foundation and have its definitive statutes approved by the government in 1900. Certain stipulations in the will had to be changed, for instance the provision that the prizes should be given for work done during the preceding year. This was obviously not practicable and instead it became possible to reward also older work in case its significance had not been established until recently. Nobel Committees of three to five members should be appointed by the awarding institutions to help with the evaluation of candidates.

To be seriously considered for a Nobel Prize the candidate must be nominated by somebody whom the awarding institution recognizes as competent. During the first decades of the 20th century this meant for physics and chemistry a few hundred individuals each year, mainly university professors and members of scientific academies in Sweden but also abroad. The number of nominations received in each subject was between 20 and 30 a year. These numbers increased slowly between the wars but after World War II the increase has been marked, not least of nominations submitted by previous laureates. For the early prizes there was of course an impressive gathering of candidates who had done very meritorious work that was now fairly old. Consequently, the question of how old a discovery could be and still be considered for a Nobel Prize was often raised and keenly debated. As we shall see in the following it could sometimes prove to be the decisive question.

The Nobel Committees elected by the awarding institutions from 1900 and onwards have carried much of the responsibility for the evaluation of the candidates. The Academy of Sciences of course had two committees, one for each of the Nobel Prizes it was entrusted with. In the physics committee Arrhenius had been a member from the very beginning until his death in 1927. Ironically enough during the first three years he had Professor Thalén as one of his colleagues on the committee. One wonders how he managed to collaborate with his old thesis advisor. We know that Arrhenius never could forget the shabby treatment his thesis had received from the Uppsala faculty. After all these years it still rankled, although he was now recognized as a leading scientist and one of the founders of physical chemistry. Nevertheless, Arrhenius rapidly became a leading member of his committee and his influence on the selection of laureates both in physics and chemistry can hardly be overrated.

Undoubtedly during the early years of the Nobel Prizes Arrhenius was much concerned with promoting the cause of the "Ionists", where the obvious prize candidates were van't Hoff, Ostwald and of course himself. One possibility was to let van't Hoff and Arrhenius share the first prize in chemistry, particularly in view of the close connection between electrolytic dissociation and osmotic pressure (van't Hoff's work on stereochemistry was considered too old for a reward). However, the Academy obviously felt that a Swedish laureate would be inappropriate at the first awarding of the Nobel Prizes in view of the great stress the donator had laid on the international character of his prizes. Furthermore, at this time Arrhenius still hoped for a prize in physics since he had always regarded himself as a physicist rather than a chemist. He therefore discreetly let it be known that he did not want a prize in 1901. However, the Swedish physicists seem to have felt that he was not really one of them and he was never seriously considered for a prize in physics. It would appear that Arrhenius from the very beginning of his studies in Uppsala had antagonized the leading physicists there and people from that group now made up the majority of the Nobel Committee for Physics. They still had a very different view from Arrhenius of what was important in physics and it is fair to say that he had more influence and better support among the members of the chemical Nobel Committee than in the physical one. The leading figure among the chemists was Per Cleve (1840–1905) who was professor of chemistry at Uppsala and had at the time voiced a very negative opinion of Arrhenius' thesis. However, he had since completely changed his mind about Arrhenius and his electrolytic dissociation theory so that he now was a great admirer and supporter of both the theory and its author. It is therefore not surprising that a leading Ionist, Jacobus van't Hoff, was the first Nobel laureate in chemistry. The physics committee had suggested the name of Conrad Röntgen to the Academy for his "discovery of the remarkable rays subsequently named after him", while the Karolinska Institutet had selected Emil von Behring for the discovery of antibodies providing protection against diptheria. Looking back on the first laureates in medicine and the sciences one must conclude that here the Nobel Prize project had come off to a magnificent start.

For the physics prize of 1902 Hendrik Lorentz and Pieter Zeeman were nominated in recognition of their work on the influence of magnetism on radiation, while the chemical prize went to one of the real giants in organic chemistry, Emil Fischer, "for his work on sugar and purine synthesis". This evened up the score between the newfangled physical chemistry and the classical organic chemistry, for so long the dominant branch of this science. In 1903 came Arrhenius' great chance for a Nobel Prize and consequently, as has always been the custom when a committee member is himself a serious candidate, he stayed away from the work of the committee. Now followed a series of almost Byzantine machinations between the chemical and physical committees. In 1902 van't Hoff had nominated Arrhenius for the prize in physics but in 1903 he nominated his fellow Ionist both in physics and chemistry. In fact, that year Arrhenius received seven nominations in physics and twelve in chemistry. His ship could truly be said to have come in. Early in 1903 the chemistry committee sent a letter to the physics committee, in those days a fairly unusual thing to happen. They pointed out that now was the time to reward Arrhenius and, furthermore, suggested a rather original way to go about it. Since Arrhenius was clearly a worthy candidate in both physics and chemistry, they argued that he should receive half of the physics prize and half of that in chemistry. The remaining half of the chemistry prize should go to William Ramsay for his discovery of the inert gases (see earlier). A prerequisite for this solution would be, however, that the other half of the physics prize should be given to Lord Rayleigh who had discovered the inert gas argon at the same time and independently of Ramsay.

The letter from the chemists inspired no enthusiasm in the physics committee. The physicists, many of them no admirers of Arrhenius, grudgingly admitted that he was a worthy candidate in physics but pointed out that there were other more deserving nominees. By that they obviously meant Henri Becquerel together with Pierre and Marie Curie for their work on radium and radioactivity. They therefore could not recommend Arrhenius for a prize in physics. As for the proposed prizes to Ramsay and Lord Rayleigh it would only be acceptable if Rayleigh got an undivided prize in physics and Ramsay one in chemistry. In other words, they wanted no joint venture by both committees that included a Nobel Prize to Arrhenius. Having thus failed to ally themselves with the physics committee, which instead went on to nominate Becquerel and the Curies, the chemistry committee now took a vote and found that three members supported Ramsay while two were in favour of Arrhenius. In the fall the physics committee, supported by the physics class (section) of the Academy, suggested the names of Becquerel and the Curies for the Nobel Prize in Physics. Meanwhile the chemistry committee was still split two-to-three between Arrhenius and Ramsay. However, when they reported to the chemistry class of the Academy the vote there was seven-to-three in favour of Arrhenius. At the plenary meeting of the Academy it was decided to award the prize in chemistry to Arrhenius and the physics prize to Becquerel and the Curies.

With the prizes to van't Hoff and Arrhenius successfully in the bag there only remained one of the leading Ionists who had not been rewarded, Wilhelm Ostwald. There could be no doubt about his role as an intellectual inspiration for the whole field of physical chemistry, where he had written seminal textbooks and started important journals, not to mention the brilliant ideas that he had lavishly scattered around. The problem was that Nobel in his will had provided for his prizes to be awarded for specific discoveries and here was a difficulty for Ostwald, in spite of his extensive publications. Where was the great discovery to be awarded? Another problem, at least for his friend Arrhenius, was the fact that Ostwald had not really adopted the concept of the atom. Instead he preferred to reason in terms of energetics, something that Arrhenius looked upon as bordering on superstition and adamantly rejected. There were also other exceedingly strong candidates. In 1904 Ramsay and Rayleigh got their prizes in chemistry and physics, respectively, and in 1905 the Academy simply had to recognize another of the German giants in organic chemistry, Adolf von Baeyer. Furthermore, this was also the year when one of the greatest of all breakthroughs in theoretical chemistry suddenly emerged in the deliberations of the Nobel Committee for Chemistry like a spectre from bygone times: the periodic law. Before we go on to consider the fate of the periodic law at the hands of the Nobel Committee, let us first establish that Ostwald eventually got his well-deserved Nobel Prize in 1909 for his work on catalysis and the principles governing chemical equilibria and rates of reaction.

An Electric Oven or the Periodic Law

In 1905 there was a change of heart in the Nobel Committee for Chemistry, which had previously argued that the frequently nominated Adolf von Baeyer could not be awarded the Nobel Prize because his great contributions to organic chemistry were too old to be considered. However, now the committee was suddenly prepared to invoke §2 of the statutes, according to which it was possible to reward older work in case its significance had not been established until recently. The Academy followed the advice of the committee and that year Adolf von Baeyer finally got his prize. Maybe it was this somewhat unexpected new attitude of the Nobel Committee that encouraged one of its leading members, Arrhenius' friend and colleague Otto Pettersson, to nominate Dmitri Mendeleev the same year, in spite of the fact that the periodic law had been suggested as early as 1869. Thus, the law had been around for at least 30 years before the Nobel Prize came into existence, which explains why Mendeleev had not been nominated before 1905. In a long and penetrating evaluation of his candidate Pettersson took pains to point out that the periodic law had become even more important because of the completely unexpected appearance of a new group (O_i present number 18) made up of the recently discovered inert gases. In fact, in the citation for the prize to Sir William Ramsay in 1904 for his discovery of these gaseous elements there was a special reference to "his determination of their place in the periodic system". A modern version of the periodic system following the International Union of Pure and Applied Chemistry (IUPAC) recommendations is presented in Fig. 26. The order of the elements is given by their atomic numbers determined by the number of positive charges (protons) in the atomic nucleus, corresponding to the number of extra-nuclear electrons. Over the years many different representations of the periodic system have been used but nowadays the form with 18 vertical groups and 7 horizontal periods shown in Fig. 26 is the most common. The elements making up a certain group all have similar chemical and physical properties since they have the same number of electrons in their outer electron shell. In previously used representations the group of inert gases, which now has the number 18, was instead called group 0. When the group of inert gases are discussed later they are referred to by their old name, group 0, since that was how they were denoted at the time when the Nobel Committee considered a Nobel Prize for Mendeleev. The seven horizontal periods are each finished by an inert gas whose outer electron shell is complete, which explains the lack of chemical reactivity of these elements. The inert gas corresponding to period 7 has still not been discovered. Throughout recent years an increasing number of artificially produced unstable elements with atomic numbers 93 and higher have been added to the periodic table. It can be noted that element 101, Mendelenium (Md) is the nearest neighbor of 102 Nobelium (No)!

Surely the time had now come to recognize with a Nobel Prize also Dmitri Mendeleev who had originally revealed this fundamental principle of chemistry to us. In their report to the Academy of 1905 the Nobel Committee selected three main candidates, of whom they unanimously recommended von Baeyer to the prize. The two other candidates seriously considered by the committee were Mendeleev and a French inorganic chemist, Henri Moissan, who had been nominated on several occasions for his isolation in pure form of the element fluorine and for his construction of an electric furnace in which very high temperatures could be obtained. In its report the Nobel Committee concluded that Mendeleev and the periodic law must be given preference over Moissan and his isolation of fluorine and construction of the electric oven. The fact that Moissan had been nominated a number of times and in 1905 had received considerably more nominations than Mendeleev (21 for Moissan against 3 for Mendeleev) did not seem to impress the committee, which was unanimous in its grading of the two candidates. Consequently, when in 1906 both Mendeleev and

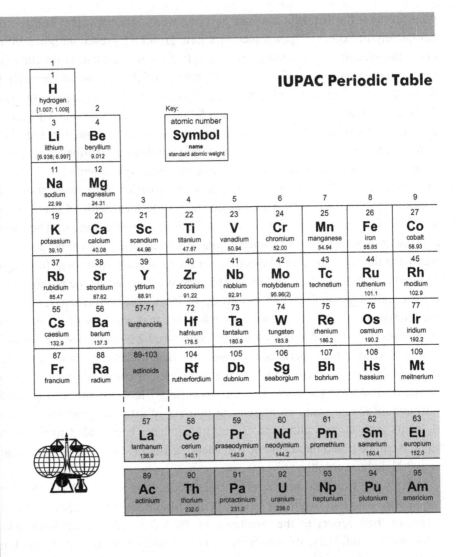

Notes

- IUPAC 2009 Standard atomic weights abridged to four significant digits (Table 4 published in *Pure Appl. Chem.* 83, 359-396 (2011); doi:10.1351/PAC-REP-10-09-14). The uncertainty in the last digit of the standard atomic weight value is listed in parentheses following the value. In the absence of parentheses, the uncertainty is one in that last digit. An interval in square brackets provides the lower and upper bounds of the standard atomic weight for that element. No values are listed for elements with no stable isotopes. See PAC for more details.

- "Aluminum" and "cesium" are commonly used alternative spellings for "aluminium" and "caesium."

For updates to this table, see iupac.org/reports/periodic_table/. This version is dated 21 January 2011.

Copyright © 2011 IUPAC, the International Union of Pure and Applied Chemistry.

Fig. 26. IUPAC 2011 version of the periodic table of the elements.

of the Elements

			13	14	15	16	17	18
								2 **He** helium 4.003
			5 **B** boron [10.80; 10.83]	6 **C** carbon [12.00; 12.02]	7 **N** nitrogen [14.00; 14.01]	8 **O** oxygen [15.99; 16.00]	9 **F** fluorine 19.00	10 **Ne** neon 20.18
10	11	12	13 **Al** aluminium 26.98	14 **Si** silicon [28.08; 28.09]	15 **P** phosphorus 30.97	16 **S** sulfur [32.05; 32.08]	17 **Cl** chlorine [35.44; 35.46]	18 **Ar** argon 39.95
28 **Ni** nickel 58.69	29 **Cu** copper 63.55	30 **Zn** zinc 65.38(2)	31 **Ga** gallium 69.72	32 **Ge** germanium 72.63	33 **As** arsenic 74.92	34 **Se** selenium 78.96(3)	35 **Br** bromine 79.90	36 **Kr** krypton 83.80
46 **Pd** palladium 106.4	47 **Ag** silver 107.9	48 **Cd** cadmium 112.4	49 **In** indium 114.8	50 **Sn** tin 118.7	51 **Sb** antimony 121.8	52 **Te** tellurium 127.6	53 **I** iodine 126.9	54 **Xe** xenon 131.3
78 **Pt** platinum 195.1	79 **Au** gold 197.0	80 **Hg** mercury 200.6	81 **Tl** thallium [204.3; 204.4]	82 **Pb** lead 207.2	83 **Bi** bismuth 209.0	84 **Po** polonium	85 **At** astatine	86 **Rn** radon
110 **Ds** darmstadtium	111 **Rg** roentgenium	112 **Cn** copernicium						

64 **Gd** gadolinium 167.3	65 **Tb** terbium 158.9	66 **Dy** dysprosium 162.5	67 **Ho** holmium 164.9	68 **Er** erbium 167.3	69 **Tm** thulium 168.9	70 **Yb** ytterbium 173.1	71 **Lu** lutetium 175.0

96 **Cm** curium	97 **Bk** berkelium	98 **Cf** californium	99 **Es** einsteinium	100 **Fm** fermium	101 **Md** mendelevium	102 **No** nobelium	103 **Lr** lawrencium

International Year of
CHEMISTRY
2011

Moissan were nominated again, the deliberations of the committee resulted in a recommendation that the Academy should award the Nobel Prize in Chemistry of 1906 to Mendeleev. However, there was one dissident who spoke eloquently and persistently in favour of Moissan.

Peter Klason (Fig. 27), professor of chemistry at the Royal Institute of Technology in Stockholm and one of the founders of Swedish cellulose chemistry, had nominated Moissan both in 1905 and 1906. Nevertheless, he had not made a formal reservation to the committee's grading of Mendeleev and Moissan in 1905, presumably because it was obvious that von Baeyer would get the prize in any case. In 1906, however, he was up in arms and ready to fight for his candidate to the bitter end. This time he officially dissociated himself from the majority of the Nobel Committee and in a separate report to the Academy he presented his arguments against Mendeleev and in favour of Moissan. The majority of the committee had stressed the importance of the discovery of the inert gases and maintained that this had added a new dimension to the periodic law so that §2 of the statutes clearly was applicable here. The full significance of the law had not been realized previously. In fact, it would seem that for every fundamental step forward in chemistry the periodic law became more important.

Against this line of argument Peter Klason in his dissident report maintained that the periodic law had for many years been part of the basic education in chemistry all over the world and was treated in all standard textbooks. Thus, the importance of the law had long been realized and consequently §2 of the statutes did not apply in this case. Furthermore, the fact that the newly discovered group 0 so elegantly fitted into the periodic system could hardly be credited to Mendeleev who had not foreseen its existence, much less reserved any room for it in his system. Finally, Klason pointed out that Mendeleev had after all based his periodic law on the principles for the determination of atomic weights first stated by Avogadro and made generally known by Stanislao Cannizzaro at the chemical congress of Karlsruhe in 1860 (see earlier). Without knowledge of the correct atomic weights the periodic system could not have been constructed. Since Cannizzaro was still alive Mendeleev could not be awarded the Nobel Prize unless the Italian chemist was also included, but he had, on the other hand, not been nominated for the prize in 1906. Further discussions by the Nobel Committee did not resolve the issue and as a result the report that eventually reached the Academy's class of chemistry was rather confusing.

For all practical purposes the decision of the Academy regarding who should get a particular Nobel Prize is based on the outcome of the deliberations of the so-called Nobel class involved, in this case the class of chemistry. The final resolution later reached by the voting of the whole Academy in plenary session is more of a formality. The problem this time was not only that the

Fig. 27. Peter Klason (1848–1937).
Courtesy of the Royal Swedish Academy of Sciences.

committee had not been able to reach a unanimous recommendation; four of its five members were in favour of Mendeleev, while one (Klason) in a dissident report advocated a prize to Moissan. There was also the problem that in the committee's report to the Academy, which incidentally was signed not only by the majority members but also by Klason and recommended Mendeleev to the prize, there was a curious finishing section in which the committee seemingly reversed its opinion or at least left room for an alternative interpretation. It went on to say roughly the following: Should the Academy agree with the minority of the committee, i.e. Klason, and be reluctant to invoke §2 of the statutes, the Nobel Committee is of the unanimous opinion that next to Mendeleev the most worthy candidate for a prize in chemistry is Professor Henri Moissan. The somewhat bewildered Academy now decided to fortify the Nobel Committee with four members of the chemical class in the hope that this reinforced body would be able to reach a decision.

If the Academy had really hoped for unanimity it was in for a disappointment. The reinforced committee was as divided as ever. During its discussions Klason appeared to give way and reluctantly admit the growing importance of the periodic law but on the other hand he insisted imperturbably that in any case Mendeleev must share the prize with Cannizzaro. Finally, he triumphantly pointed out that Cannizzaro had not been nominated for 1906 and consequently could not be a candidate that year. He then resumed his indefatigable arguments in favour of Moissan and in the end he was able to persuade a majority of the fortified committee, which eventually voted five to four for Moissan. Two of its members, one in favour of Moissan and the other for Mendeleev, submitted memos to the Academy explaining the reason for their respective votes. In the end the Academy in plenary session decided to award the Nobel Prize in Chemistry for 1906 to Henri Moissan.

It might be of interest to examine more closely the arguments presented in the memos of the committee members in explanation of their votes. They were both distinguished scientists. Count Karl A. H. Mörner was professor of medical biochemistry and president of the Karolinska Institutet while Olof Hammarsten was professor of medical and physiological chemistry at Uppsala University and the leading biochemist in Sweden. He was also president of the university and a member and later chairman of the chemical Nobel Committee. The two memos are indeed very different in their general attitude. Count Mörner takes a very bureaucratic and formalistic line in his memo and is mainly interested in the question of whether §2 of the statutes can be applied in the case of Mendeleev and the periodic law. He gives great weight to the fact that the law has been known and appreciated since 1869 and points out that Mendeleev was given the Davy Medal in 1882. He is also unwilling to give Mendeleev any credit for the fact that the inert gases of group 0 fit so nicely into the periodic system since Mendeleev did not foresee them nor reserve room for them in his system. On the whole he seems to agree with Peter Klason's view and his punch line is that in spite of his high regard for Mendeleev he must conclude that the wording of the statutes as well as their spirit makes it impossible for him to accede to the suggestion that Mendeleev should be awarded the Nobel Prize in Chemistry. Perhaps the most surprising thing about this line of reasoning is that the year before the prize had been given to Adolf von Baeyer in spite of the fact that his great contributions to organic chemistry had previously on several occasions been considered too old for a prize.

Hammarsten begins his memo by emphasizing that it is imperative to distinguish between the scientific importance of the candidates' research and the formal question of whether §2 of the statutes is applicable in the present case. He then goes on to say that in his opinion Mendeleev of all the candidates nominated has the greatest research merits for the prize. From a purely scientific point of view he is in a class by himself. As far as §2 and the inert gases are concerned Hammarsten takes the same view as Klason and Count Mörner and agrees with them that Mendeleev had no part in this new discovery. On the other hand, he draws an entirely different conclusion from this opinion. It is precisely the fact that Mendeleev had *not* foreseen group 0 and had made no room for it in his system that demonstrates the ability of the periodic law to explain previously unexpected qualities of the elements. This is indeed a case of what §2 calls older work whose significance has not been established until recently. Hammarsten therefore insists on his previously expressed opinion that Mendeleev and his periodic law are eminently qualified for a Nobel Prize in Chemistry.

The Nobel Committees have often enough witnessed heated discussions, not least concerning the interpretation of §2 of the statutes. The great German bacteriologist Robert Koch, who had after all together with Louis Pasteur founded a new science, microbiology, had experienced considerable difficulties with §2 before he finally got his prize in 1905. Nevertheless, the case of Mendeleev and the periodic law is perhaps the most notorious of them all. One reason for this is of course that he died early in 1907 and thus left the Academy without the opportunity of later correcting its mistake. Because a serious mistake it certainly was. On the face of it the interpretation of §2 would seem to dominate the discussion, but is that really all? Peter Klason had written two almost identical evaluations of Moissan for the committee and they both have the same introductory remarks: "Henri Moissan belongs to those contemporary chemists who are better known for their experimental and technical skill than because of their ability to come up with original and seminal ideas in science." In our eyes this is a curious way to begin an evaluation which eventually leads to an enthusiastic recommendation for a Nobel Prize. We would be more inclined to think that Klason is here pointing out the lack of certain abilities in Moissan's scientific talents, which we would consider particularly important for a Nobel laureate to possess.

Today it goes without saying that the periodic law is perhaps the most decisive progress ever made in theoretical chemistry, but this may not have seemed such a merit to Klason who with his practical-industrial background obviously attached more importance to the construction of an electric oven. In fact, he was not the only one to come to a similar conclusion. After all, Moissan had received many more nominations than Mendeleev both in 1905 and 1906. Nor was it just a question of chauvinistic French colleagues nominating a fellow countryman. In the period 1905–1906 his supporters included such names as Emil Fischer, Adolf von Baeyer and William Ramsay. In his letter of nomination Sir William Ramsay also "ventured to state in reply to the argument that the nobel gases are a confirmation of Mendeleev's periodic system that we are all using Cannizzaro's revolution in chemical ideas and Cannizzaro is still alive". Of course, Klason was happy to use Ramsay's opinion in his own arguments with the majority of the Nobel Committee.

We should also remember that many colleagues in Russia had criticized Mendeleev for being inclined to unsupported speculations. Generally speaking, in the chemistry of those days, untiring experimental work at the bench may have been regarded as more meriting than high-flying theories. There were several exceptions, of course, in particular Otto Pettersson who was a very strong supporter of Mendeleev and had twice nominated him for the prize. One might have thought that Pettersson's friend Svante Arrhenius, who was very influential in the Academy since he was their only Nobel laureate, would also have supported Mendeleev. However, Mendeleev had been less than enthusiastic about Arrhenius' theory of electrolytic dissociation, which did not fit in with Mendeleev's general ideas about chemistry. Criticism of that theory was always a sensitive issue for Arrhenius and he may have had difficulties overlooking the lack of support from Mendeleev. Furthermore, Arrhenius was becoming increasingly fascinated by the work of Ernest Rutherford and his theories about the disintegration of radioactive elements. Here was indeed a candidate who got a Nobel Prize in 1908, which Arrhenius very actively promoted. However, he does not seem to have been in favour of a prize to Mendeleev.

The day of 27 October, when the class of chemistry of the Royal Swedish Academy of Sciences voted five to four in favour of a Nobel Prize to Henri Moissan (Fig. 28) and against Dmitri Mendeleev, is not one of its more glorious days. One might even call it a nadir of prestige. Whom should we blame? It is tempting to make Professor Peter Klason our scapegoat. He had after all done his

Fig. 28.   Henri Moissan (1852–1907).
From Les Prix Nobel en 1906.

utmost to bring about the catastrophe. On the other hand, a number of the great chemists of the day had agreed with him. Moissan was solid and dependable while Mendeleev was inclined to speculate in much the same way that had been the bane of chemistry for centuries before we learned to think scientifically and critically. Perhaps it was the millennia of chemical history that caught up with Dmitri Mendeleev so that he fell victim to the memory of such figures as Paracelsus and the alchemists with their unbridled fantasies of magic potions and the philosopher's stone.

It is in the nature of the Nobel Prize that there will always be a number of candidates who obviously deserve to be rewarded but never get the accolade they hoped for. Of course it is an impossible task that the Academy has struggled with for more than a century. Maybe one can claim that it has done a fairly good job after all; it has certainly tried very hard. Nevertheless, there have been unfortunate sins of omission and the most outrageous is probably that committed against Dmitri Mendeleev. It is indeed a pity that his name does not appear in the distinguished list of laureates. It would certainly have added to the prestige of both the Nobel Prize and the Royal Swedish Academy of Sciences.

Bibliography

Arrhenius, Svante. Über die Dissociation der in Wasser gelösten Stoffe. *Z. Phys. Chem.* **1**, 631–648 (1887).

Brauner, Bohuslav. Contributions to the Chemistry of Rare Earth-Metals. *J. Chem. Soc. Trans.* **41**, 68–79 (1882).

Cannizzaro, Stanislao. Sunto di un corso di filosofia chimica fatto nella Reale Università di Genova. *Nuovo Cimento* **7**, 321–366 (1910).

Crawford, Elisabeth. *The Beginning of the Nobel Institution: The Science Prizes, 1901–1915* (Cambridge University Press, Cambridge, 1984).

Danzer, Klaus, Dmitri I. Mendelejew and Lothar Meyer. *Die Schöpfer des Periodensystems der chemischen Elemente* (B.G. Teubner Verlagsgesellschaft, Leipzig, 1974).

Frängsmyr, Tore (ed.). *Science in Sweden. The Royal Swedish Academy of Sciences 1739–1989* (Science History Publications, Canton, M.A., 1989).

Friedman, Robert Marc. *The Politics of Excellence: Behind the Nobel Prize in Science* (W.H. Freeman, New York, 2001).

Fruton, Joseph S. *Molecules and Life* (Wiley-Interscience, New York, 1972).

Gay-Lussac, Joseph Louis. Memoire sur la combinasion des substances, les unes avec les autres. *Mem. Phys. et de Chim. de la Soc. d'Arcueil* **2**, 207–234 (1809).

Gordin, Michael D. *A Well-Ordered Thing, Dmitri Mendeleev and the Shadow of the Periodic Table* (Basic Books, New York, 2004).

Hammarsten, Olof. Zur Kentnis der Nukleoproteide. *Z. Physiol. Chem.* **19**, 19–37 (1894).

Jorpes, Erik. *Jacob Berzelius: His Life and Work* (Almquist and Wiksell, Stockholm, 1970).

Koch, Robert. Verfahren zur Untersuchung, zum konservieren und photographieren der Bakterien. *Beitr. Biol. Pflanz.* **2**, 399–434 (1877).

Lagerkvist, Ulf. *DNA Pioneers and Their Legacy* (Yale University Press, New Haven, 1998).

Lagerkvist, Ulf. *Karolinska Institutet and the Struggle against the Universities* (Gidlunds Förlag, Hedemora, Sweden, 1999) (in Swedish).

Lagerkvist, Ulf. *Pioneers of Microbiology and the Nobel Prize* (World Scientific, Singapore, 2003).

Lagerkvist, Ulf. *The Enigma of Ferment* (World Scientific, Singapore, 2005).

McKie, Douglas. *Antoine Lavoisier* (Constable, London, 1952).

Melhado, Evan M. and Frängsmyr, Tore (eds.). *Enlightenment Science in the Romantic Era* (Cambridge University Press, Cambridge, 1992).

Mendeleev, Dmitri. Zusammenhang zwischen den Eigenschaften und dem Atomgewicht der Elemente. *Ber. Dtsch. Chem. Ges.* **2**, 553 (1869).

Mendeleev, Dmitri. Über das natürliche System der Elemente und seine Anwendung zum Ermitteln der Eigenschaften unentdeckter Elemente. *Ber. Dtsch. Chem. Ges.* **3**, 990–991 (1870).

Mendeleev, Dmitri. Zur Frage über das System der Elemente. *Ber. Dtsch. Chem. Ges.* **4**, 343–352 (1871).

Mendeleev, Dmitri. Die periodische Gesetzmäsigkeit der chemischen Elemente. *Liebigs Ann.*, Suppl. **8**, 133–229 (1871).

Meyer, Lothar. Die Natur der chemischen Elemente als Funktion ihrer Atomgewichte. *Liebigs Ann.*, Suppl. **7**, 354–364 (1870).

Moissan, Henri. Action d'un courant électrique sur l'acide fluorhydrique anhydre. *C. R. Acad. Sci.* **102**, 1543–1544 (1886).

Nilson, Lars Fredrik. Sur l'ytterbine, terre nouvelle de M. Marignac. *C. R. Acad. Sci.* **88**, 642–647 (1879).

Ostwald, Wilhelm. *Lehrbuch der allgemeinen chemie* (W. Engelmann, Leipzig, 1885).

Partington, James Riddick. *History of Chemistry* (Macmillan, London, 1961–1970).

Patterson, Elizabeth C. *John Dalton and the Atomic Theory* (Doubleday Co., New York, 1970).

Schück, Henrik, Sohlman, Ragnar, Österling, Anders, Liljestrand, Göran, Westgren, Arne, Siegbahn, Kai, Siegbahn, Manne, Schou, August and Ståle, Nils. *Nobel, The Man and His Prizes* (American Elsevier Publishing Co., New York, 1972).

Strutt, John William, Baron Rayleigh and Ramsay, William. Argon, a New Constituent of the Atmosphere. *Phil. Trans. R. Soc. Lond. A* **186**, 187–241 (1895).

Tansjö, Levi. *From Lavoisier to Strindberg. Lectures in Chemical History* (Berzelius Sällskapet, Lund, Sweden, 2008) (in Swedish).

Taylor, Frank Sherwood. *Alchemists, Founders of Modern Chemistry* (Henry Schuman, New York, 1949).

van 't Hoff, Jacobus Henricus. Die Rolle des osmotischen Druckes in der Analogie zwischen Lösungen und Gasen. *Z. Phys. Chem.* **1**, 481–508 (1887).

Index